The Width of the Present

Eduardo Padilla-Diaz

The Width of the Present

Book 2 from the series: Modeling the Body-Mind

Eduardo Padilla-Diaz

Eduardo Padilla-Diaz

Table of Contents

0. [Cover-2] *The Width of the Present - Book 2 of the series: Modeling the Body-Mind.*
1. [TOC] *Thematics and Exploration Plan.*
2. [Conventions] *Considerations about revisions, references, TOCs, and Epilogs.*
3. [Abstract 2] *Quantum Thinking Physics is the discrete information framework consciousness uses.*
4. [Serialization] *The most crucial transformer in understanding and exchanging our thoughts.*
5. [Data Structures] *Matter dwells in an undetermined probability state without structuring.*
6. [Width of the Present] *Paradox used by the conscience to insert itself in the domain of time.*
7. [Time Segments] *They only exist in the mind to give us access to the present, past, and future.*
8. [BioBots] *These wonderful beings inhabit our body from the molecular to the multi-organic rings.*
9. [Molecular Nets] *The intracellular networks where molecular robots operate.*
10. [Neural Nets] *Neurons connect reality, virtuality, and modulators with body and mind so we can be.*
11. [Bot Nets] *Networks where cellular robots circulate to transport supplies, repair, and protect.*
12. [Signaling] *Signals used in the neural and cellular networks to exert actions in cells.*
13. [Memories-2] *These operational memories store the innate, immediate, and acquired knowledge.*
14. [Knowledge] *It is where we store what we know to help us decide what to do next.*
15. [Issues] *Gradual conscient activities processed by thinkers at the thinking center.*
16. [Channels] *Stream useful information from sensory centers to thinkers and motor centers.*

17. [Entanglement] *How the object of interest under focus entangles with thinking and consciousness.*
18. [Quantumness] *The fabric of quantum thinking physics: a paradigm for modeling consciousness.*
19. [VC-2] *Visual Center 2: Automatic detection of object attributes, reflexes handling, and more.*
20. [MC-2] *Motor Center 2: Locomotion, balance, movement fuidity, precision, reflexes, command interface.*
21. [Thinking Messages] *Sensory informatics paradigm bridges changes and states to the thinking system.*
22. [Thinking-2] *Quantum thinking in all thinkers, attention arbitration, anticipacion, and more.*
23. [Awareness] *Is the state of conscious comprehension of all thinking and feeling of all sensations.*
24. [Epilog-2] *Thematics of the next book in the series: Tokens and Words.*
25. [Copyrights] *List of copyright certificates that protect these works.*
26. [Dedication] *To my children: Mariana, Andres, and Camilo.*
27. [Back-cover-2] *Brief about the author Eduardo Padilla-Diaz and his works.*

© Eduardo Padilla-Diaz - All Rights Reserved. epadilladiaz@gmail.com

Eduardo Padilla-Diaz

Conventions

"Considerations about revisions, references, TOCs, and Epilogs."

2.1 Introduction

- In this work, the following conventions are followed:
1. Most of the paragraphs in the book are marked with icons that contain metadata (such as edition dates); at the reader level, they have the following meaning:
    - Final edition paragraph.
    - The paragraph contains a concept, proposal, or idea originated by the author Eduardo Padilla or a statement of something already existing but explained by the author in his own words.
2. Wikipedia is the preferred reference source throughout this book since it provides access to all references at no cost or membership ties. Furthermore, Wikipedia almost always extends to more authoritative reference sources. Since its creation in January 2001, Wikipedia has become the world's largest reference website, attracting more than one billion visitors monthly. It currently has more than fifty-nine million articles in more than 300 languages.
    - [wikipedia] wikipedia
3. The Tables of Contents and the Epilogs of every book of the series form a summary of the explorations.

© Eduardo Padilla-Diaz - All Rights Reserved. epadilladiaz@gmail.com

Abstract 2

※ *"Quantum Thinking Physics is the discrete information framework consciousness uses."*

3.1 Introduction

- ◆ Reflections on this work:
- ※◆ This book explores how the mind handles space-time to achieve our human mind's extraordinary consciousness. I present several *paradigms* and hypotheses, including: a) The width of the present, a *paradox* the conscience uses to insert itself in the domain of time. b) Time segments that only exist in the mind to give us access to the present, past, and future. c) The entanglement of the object of interest under focus that leads to the awareness of its state by the conscience. d) We'll delve into quantum concepts, from sensations to awareness, under our mind's linguistic framework, where thinking and consciousness occur; I call this framework *Quantum Thinking Physics*. e) Quantumness is a space-time *paradigm* the mind uses to handle efficiently the discrete information processing needed by consciousness coherent comprehension of all thoughts and awareness superposition of all feelings. f) Quantum thinking is an information fragmentation *paradigm* required to efficiently handle the concurrent processing of the multiple *thinkers* needed to fulfill consciousness' desire to process all outstanding issues in the thinking pool quickly and independently—according to their priority.
  - ○ [space-time] [paradigm] [paradox] wikipedia

- 🛈 Compared to the previous book, The Rings of the Mind, this book presents an improved version of the thinking system, including: a) A Thinking Center that allows several *thinkers* to process each a thought independently. b) The *thinkers* are *event-driven*, making them more effective in the time domain. c) Awareness is included in the models and shows how it connects to the sensory, motor, and thinking. d) A Motor Center that can process several motor functions simultaneously and independently. e) The Motor center forms a *closed-loop controller* with the Visual Center to automatically maintain the object of interest under focus with little or no intervention from thinking and consciousness. f) A Visual Attention Center is included in the model to arbitrate requests from the sensory, thinking, and consciousness. g) The Visual Center at the operational level facilitates its interface with conscience, the Visual Attention Center, and the Head Motor Center. h) The Visual Center is improved at a cognitive level, compensating for inevitable delays and guaranteeing cognitive coherence, supporting prediction and anticipation.
    - [event-driven] [closed-loop controller] wikipedia
- 🛈 This work is composed, so far, of a series of four books: The Rings of the Mind, The Width of the Present, Tokens and Words, and The Domains of the Conscience. The work progressively presents the fundamental knowledge needed to understand and comprehend the models of thinking.

(c) Eduardo Padilla-Diaz - All Rights Reserved. epadilladiaz@gmail.com

Serialization

"The most crucial transformer in understanding and exchanging our thoughts."

4.1 Introduction

- Reflections on *Serialization* (part 1):
- *T* (*Thinker* artificial framework) is a parallel external world somewhat similar to our mind. It is very primitive but still has enough functionality that I use to show clear examples of how our mind would implement certain functions. You will understand what I mean as I refer to *T* throughout these books.
    - *T* is a hardware/software set of tools I use to assist me in modularizing and modeling the different functional components of our body-mind.
    - *T* benefits the reader's understanding and comprehension of the body-mind components postulated across this work.
- *Serialization* is an essential concept in the exchange of information. It occurs in visualization, speech, writing, data storage, and communications.
    - *Serialization*: in general terms, is the process of converting objects into a sequence of symbols suitable for:
        - Exporting objects to external minds.
        - Importing objects from external minds.

- Transporting objects across connections.
- Saving objects on artificial permanent storage.
- Retrieving objects from artificial permanent storage.
- This chapter covers the following concepts:
  - Language *Serialization*
  - T's Object *Serialization*
  - *Serialization* Distortion

4.2 Language Serialization

- Language is a *Serialization* mechanism that allows us to export thoughts using *Tokens*.
  - When we write, we serialize our thoughts with word identifiers called *Tokens* and transfer them to paper as *Token*-based thoughts.
  - When we speak, we serialize our thoughts with word identifiers called *Tokens* and transfer them to air *Token*-based thoughts.
- Language is also a de-serialization mechanism that allows us to import thoughts using *Tokens*.
  - When we read, we read from paper: *Tokens*, and our brain converts those *Tokens* into words that recreate the thoughts described on paper.
  - When we listen, we hear *Tokens*, and our brain converts them into words that recreate the thoughts described by the speaker.

4.3 T's Object Serialization

- This section introduces the concept of *Object Serialization*[1] under the T environment:
  - In T, *Object Serialization*[1] is the process of converting an object into a text description suitable for:
    - Storing it in the associative memory.
    - Retrieving it from the associative memory.
    - Transporting it across a network.
    - A *Descriptor* is the string resulting from the *Serialization* of an object.
    - A *Clone* is an object created from the *Descriptor* of an object.
    - In T, *De-Serialization* is the process of converting a text description of an object into a *Clone* of the original object.
    - In T, objects are serialized using *XML*[2], an object encoding standard suitable for serializing complex structures of objects.
  - To clarify the above concepts under this context, let's consider the following object, which is an image of a yellow circle with a green border:

Fig. 4.1

- ◉ The previuos image object is serialized using the following three lines of *XML*:

```
<svg width="100" height="100" >
  <circle cx="50" cy="50" r="50" stroke="green" stroke-width="2" fill="yellow">
</svg>
```

Fig. 4.2

- ◉ Let's analyze the above XML line by line:
    - ◉ The 1st line declares the object as an SVG (Scalar Vector Graphics) image; this object's width and height attributes (100x100) determine the size in pixels of the canvas where the image is drawn.
    - ◉ The second line declares a circle object positioned in the canvas at the coordinates cx=50 and cy=50, with a radius 50. It is stroked green and filled with yellow.
    - ◉ The third line declares the end of the SVG object, which acts as the container of the circle object. Under the SVG environment, objects can contain objects deep to any level.
- ◉ The following figure shows the model of how an XML declaration of a circle is rendered by the XML To Image Converter device; it also shows how the circle image can be converted into XML by the Image to XML Converter.

---

1. [Serialization] [XML] [SVG - Scalable Vector Graphics] [JSON] [ASCII] wikipedia

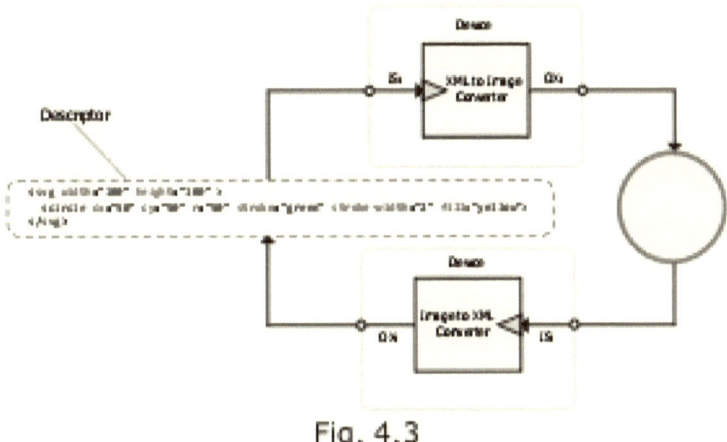

Fig. 4.3

4.4 Serialization Distortion

- ⊕ *Serialization Distortion* measures how much a *Clone* differs from the original.

- ⊕ To clarify this concept, let's consider the following object, which is an image of a yellow circle with a green border:

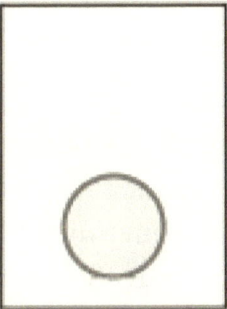

Fig. 4.4

- ⊕ Let's assume that the above figure is printed on the cover of a book, and we want to describe the cover in writing; we might describe it as follows:

- *"The book cover has a yellow circle, and the circle has a green border."*
- Suppose the above cover description is read by two individuals who have never seen the cover before. In that case, they might interpret the cover description as shown on the following covers:

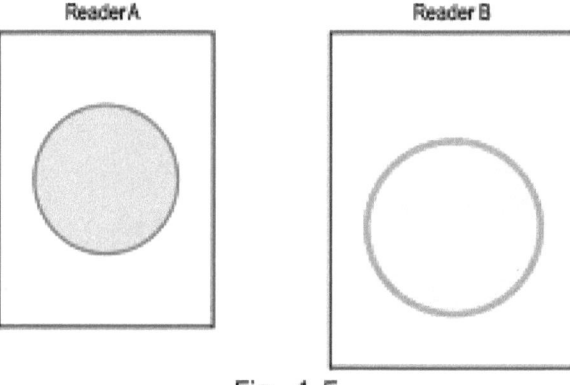

Fig. 4.5

- Notice that the interpretations of the cover description are valid for both readers because both match the description; however, the interpretations of the readers differ from the original because the description lacks the necessary information to describe the original cover better.
- Consider this case: If the book is on display in a bookstore and the two readers of our example are asked to find it in the store, they will probably find it. The readers might realize that the book in the store differs from the one they had in mind, but both readers will be able to recognize the book they found as the one they were looking for.
  - The fidelity measures how much a Clone differs from the original object.

- ⊕ The original description must include additional information to match the cover description with the actual cover.
  - ⊕ Our language has words that play essential roles required to obtain a higher level of fidelity when we describe object states; for example, prepositions locate objects in the space and time domains, and adjectives describe their matter physical properties.

  © Eduardo Padilla-Diaz - All Rights Reserved. epadilladiaz@gmail.com

Eduardo Padilla-Diaz

Data Structures

"Matter dwells in an undetermined probability state without structuring."

5.1 Introduction

- ◉ Reflections on Informatics (part 1):
- ◉ Informatics is the scientific study of information representation, processing, and communication.
  - [informatics] wikipedia
    - ◉ Computer Science is the study of computation, information, and automation.
      - [computer science] wikipedia
- ◉ Here, we will explore an important aspect of information processing: *Data Structures*.
- ◉ A *Data Structure* is a data organization, management, and storage format chosen for efficient data access.
  - [data structure] wikipedia
- ◉ *Data Structures* are classified into several types, but for now, we are going to introduce two of the most important: *Queue* and *Tree*:
  1. ◉ A *Queue* is a structure where items are placed in order of arrival to process them on a first-come-first-served basis, similar to a bank's waiting line.
     - [queue] wikipedia

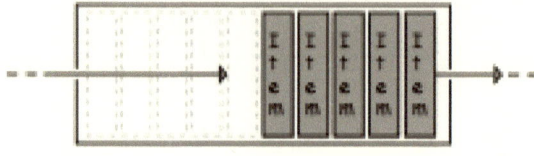
Fig. 5.1

- ⚙ In our body, one of the clearest examples of *Queues* is found in the input section of an acoustic thinker. There, the speaker's word identifiers of a sentence are temporarily stored in an understanding *Queue* (uQ) as they are received.

  - ⚙ The following figure shows the content of the uQ after queuing the sentence, 'My dog has a tail.'
    - [queue] wikipedia

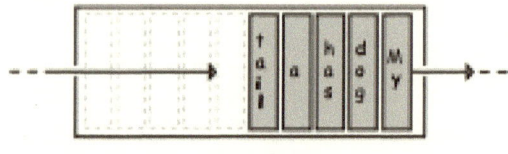
Fig. 5.2

- ⚙ When a word identifier is added to the uQ, the Acoustic Center (AC) searches the acoustic vocabulary for its semantic value to find its meaning and copies it to a comprehension *Queue* (cQ). At the end of the sentence, the AC passes the cQ and control to the comprehension section, where it is analyzed with the help of the conscience, which will decide what to do next. The uQ is cleaned between sentences.

- ⚙ One important thing to remember: The Magical Number Seven, Plus or Minus Two, is one of psychology's most highly cited papers. It was written by the cognitive psychologist George A. Miller of

Harvard University's Department of Psychology and published in 1956 in Psychological Review. It is often interpreted to argue that the number of items an average human can hold in short-term memory is 7 ± 2; this has occasionally been referred to as Miller's law.

- [7 +/- 2] wikipedia

2. ⁕ A *Tree* is a *Data Structure* that stores data associated hierarchically, similar to the computer program File Explorer, shown in the following figure:
   - [tree] wikipedia

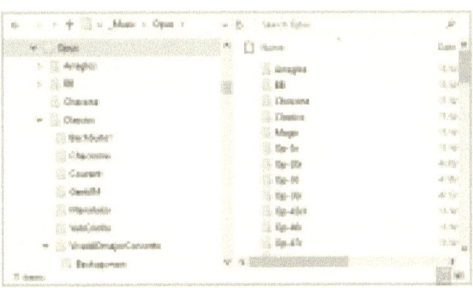

Fig. 5.3

- ⁕ One of the clearest examples of using *Trees* in our bodies is storing and retrieving acquired semantic knowledge.

  - ⁕ For example, when a *thinker* processes the sentence: *My dog has a tail*. In the execution phase, the *thinker* adds or updates the *tail* in the 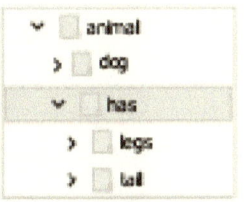 semantic vocabulary at the 'has' branch of the dog. The *tail* is used to retrieve the semantic knowledge of the word *tail*.

© Eduardo Padilla-Diaz - All Rights Reserved. epadilladiaz@gmail.com

Eduardo Padilla-Diaz

Width of the Present

① *"Paradox used by the conscience to insert itself in the domain of time."*

6.1 Introduction

- ⊕ Reflections on time (part 1 of 3):
- ⊕ Imagine we are traveling on a train on a lunar eclipse night. The wagon that transports us has no windows. The only way to see the visual reality of the outside is through photos taken with a periscope that contains a particular type of Polaroid camera.
    - [polaroid camera] wikipedia
    - ⊕ The moon will be our main object of interest, and we will focus on it when we take the photos.
- ⊕ When you take a photo with a Polaroid camera, you have to wait around 10 seconds for the camera to develop and print it on photographic paper.

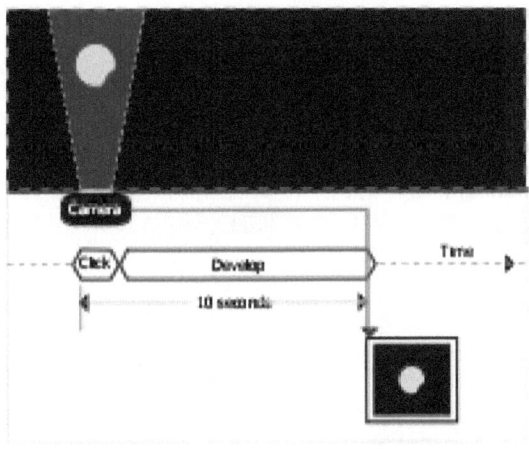

Fig. 6.1

- ○ ◉ As soon as the camera prints a photo, it represents the present of the external visual reality ten seconds ago.
- ◉ Being in the wagon, if we want to continuously inform ourselves of the external visual reality, we must take a photo every 10 seconds. Based on this, we can then say the following:
  - ○ ◉ The present we see through a series of newly printed photos is always 10 seconds delayed from the actual visual reality.
- ◉ In the case of the train, we can establish the following:
  - ○ ◉ There is a delay between the excitation caused by external visual reality and the conscious perception of it.
- ◉ Now let's consider that - in the case of the mind- conscious visual perception behaves similarly to the case of the train:
  - ○ ◉ There is a delay in the Visual Center, which is a few milliseconds, for our mind to consciously perceive the external excitation of visual reality.
  - ○ ◉ In general, there is a delay between the excitation caused by external physical reality and its conscious perception in each of our senses.
  - ○ ◉ Later, we will see something interesting: it will show us that there exists in us a reflex sensory perception, which is much faster than conscious sensory perception, which, under certain conditions,

automatically controls our motor part, and which is primarily aimed at protecting the integrity of our body and mind.

6.2 The Width of the Present

- ⊕ Let's continue with the case of the train and refer to the following figure:

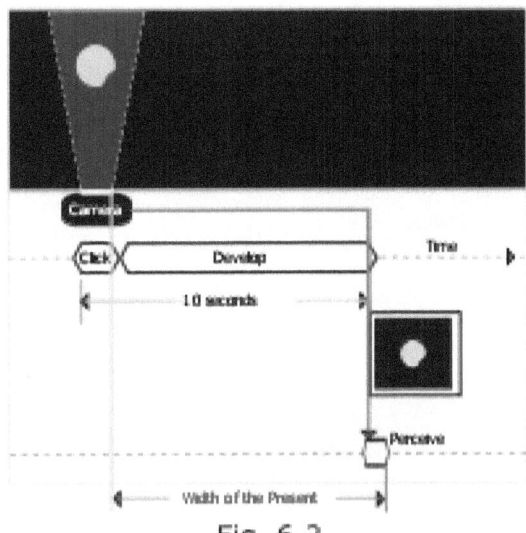

Fig. 6.2

- ⊕ When you trigger the camera to take the photo, the following happens:
  1. ⊕ The camera opens the shutter, and the light excites its optical sensor.
  2. ⊕ The shutter closes, starting the width of the present (green vertical line).
  3. ⊕ At this point, the optical sensor contains a copy of the visual present.

4. ⦿ The camera processes the data from the optical sensor, converting it into a format suitable for printing on photosensitive paper.
5. ⦿ The photo is printed.
6. ⦿ If we focus on the photo, we see it, consciously perceive it, and mark this point as the end of the width of the present (purple vertical line).
    - ⦿ But, if we do not focus on the photo, we will not perceive it, and it is as if we did not see it.

- ⦿ In the case of the train, we can then say:
  - ⦿ The width of the present is the time interval between the optical exposure to the visual reality and the delay in perceiving the photo consciously.

- ⦿ In the case of the mind, we can say that the brain, with its sensory organs, such as, among others, the eyes and ears, with their respective sensory centers, where the excitations of physical reality are processed, there is also a width of the present for each of them:
  - ⦿ In the case of our eyes, the width of the visual present is the time interval between exposure to visual reality and the delay in perceiving it consciously.
    - ⦿ In the visual center, the delay between the exposure of the image and its perception is on the order of 100 milliseconds.
  - ⦿ In the case of our ears, the width of the acoustic present is the time interval between exposure to acoustic reality and the delay in perceiving it consciously.

- ◉ In general, in each sensory center's case, the width of the present is the time interval between exposure to reality and the delay in perceiving it consciously.

- ◉ The width of the present is not concurrent; it only occurs in the sensory center and Thinker, which are processing the object of interest in which our conscience has the focus.

- ◉ Our conscience can rotate the focus among several issues the mind keeps on a list, which we will call the attention list. The attention list is dynamic in terms of the number of issues and the priority assigned to each. The focus is given to the list's member with the highest priority, which is determined by the conscience. The priority of each issue in the list can vary from moment to moment and is given by the conscience according to the results of the last perception. Thinking, attention, and conscience are discussed later in their chapters.

6.3 Visual Reality Queue

- ◉ Remember that a queue is a structure where objects are placed in order of arrival to process them on a first-come-first-served basis, similar to the waiting line of a bank.

- ◉ Every time we take a photo of the eclipse, we put it first in the visual reality queue, which can only hold a limited number of photos. When the queue is full, the oldest photo is removed to make room for the most recent one. The time a photo remains in this queue is therefore limited.

- ⊕ On the other hand, if we consider a photo in the visual reality queue useful and meaningful, we place it in another queue called the visual comprehension queue, where it will be part of a new visual record.

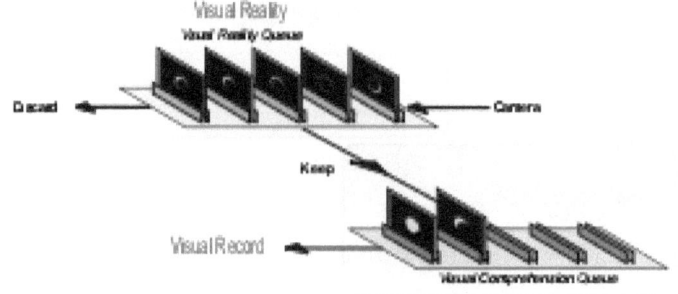

Fig. 6.3

- ⊕ We fill the visual comprehension queue only with useful and meaningful photos. We do not want to fill visual records with hundreds of useless photos because their slots must be dedicated to storing useful information.
- The visual comprehension queue helps us extract meaning from our perceptions to build a useful and understandable visual record.

6.4 Data Structures

- ⊕ In the case of the train, the comprehension memory is a temporal, visual, mechanical memory with a queue structure.
- ⊕ In the case of the mind, the comprehension memory is a temporary, visual, organic memory with a queue structure. The records of visual memories are

permanent, visual, organic memories with a tree structure

6.5 Immediate Visual Memory

- ◉ In the case of the train, the visual comprehension queue is an immediate visual memory where the most recent significant information about visual reality is stored in order to understand it.
- ◉ In the case of the mind, an immediate visual memory is nothing more than something similar to the visual comprehension queue described above.

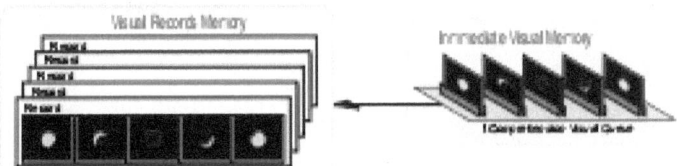

Fig. 6.4

- ◉ The immediate visual memory is a temporary memory that helps us extract meaning from a sequence of relevant images to build a useful, meaningful, and understandable visual record.
- ◉ The memory of visual records is a permanent memory, where we store all types of visual records; where each visual record is useful, meaningful, and understandable; it is stored in chronological order; can have a different number of images in sequence; it may be part of other sequences of visual records.

Eduardo Padilla-Diaz

The Width of the Present

Time Segments

🕐 *"They only exist in the mind to give us access to the present, past, and future."*

7.1 Introduction

- 🌕 Reflections on time (part 2 of 3):
- 🕐🌕 In the mind, there are seven main time segments: present, immediate past, immediate future, near past, near future, past, and future. The following figure shows a model of the time segments:

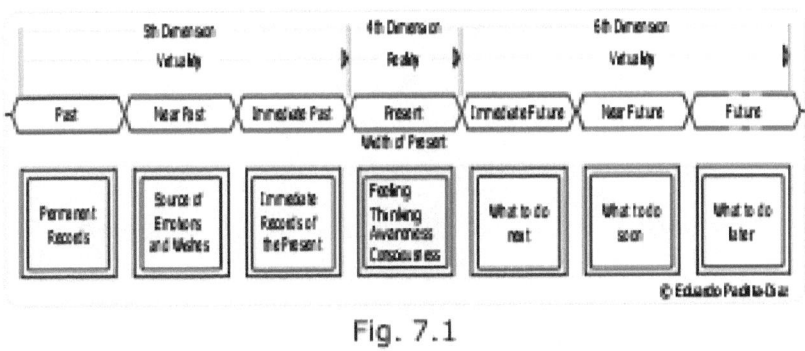

Fig. 7.1

1. 🕐🌕 In the present, time is real; in all other segments, time is virtual.

    - 🕐🌕 The segment of the present is the most important of all; in it, we perceive reality, protect ourselves, think, and operate.

    - 🕐🌕 The segment of the present is where everything takes place: thinking, remembering, imagining, dreaming, all of which occur in the present. When we dream, time and space are virtual in the context of the dream, but the synthesis of the dream takes place in the present.

2. In the immediate past, time is virtual.

    - The immediate past is the second most important. In chronological order, this segment contains the most relevant of what was done in the last few presents. The information in this segment is essential for understanding language, anticipating, protecting ourselves, and many other things.

3. In the immediate future, time is virtual.

    - The immediate future segment is the third in importance. This segment contains a list, in order of importance, of what we have to do in the next few presents. The priority of list members may be changed, and list members may be removed at any time. The information in this segment largely determines our behavior.

    - The immediate future determines what we will do after we finish what we are doing in the width of the present.

    - The immediate future can be called - with very high probability - the present of the future since it contains what we will do immediately. Only an unexpected event - like a reflex action - could change what will be done in this segment.

    - We can say with high certainty that the present of the future is predictable from the point of view of our behavior.

4. 🕐 In the near past, time is virtual.

    - 🕐 The segment of the near past is the fourth in importance.
    - 🕐 Various types of knowledge are stored in the near past: what we have done, what we have learned, what we do not want to forget, and what we should avoid.
    - 🕐 The near past determines part of what we are going to do in the future, what we want to know, what we should know, what we want to learn, what we must learn, fulfill what we promised, what we want to change, the experiences that we want to repeat.
    - 🕐 The near past plays a very important role in the emotional part of acquired tastes: in our hobbies, in love, in the places we want to visit, in the people we want to see again, in the people we should see again, and more.
    - 🕐 The near past is the time segment, emotional par excellence.
    - 🕐 The near past works in concert with the near future attention list, as all the desires that emerge from our experiences in the near past are added to the list.

5. 🕐 In the near future, time is virtual.

    - 🕐 The segment of the near future is the fifth in importance; it works for the needs we must solve soon. One form of attention, called near-future attention, is part of the episodic acquired knowledge of the near future and contains a list of needs that must be attended to soon. The content of this segment determines our functional behavior.

6. In the past, time is virtual.

    - The segment of the past is the sixth in importance. This segment contains what we remember about what happened, what we remember about what we did, and what we remember about what we would have done. The information in this segment is essential for knowing the environment and exploring the habitat, remembering where there is food and danger, understanding the meaning of words, and many other things.

7. In the future, time is virtual.

    - The segment of the future is the seventh most important. This segment contains forecasts of everything people tell us will happen, everything we will do, and everything we would like to do. This information is essential for the exploration of new habitats, the search for new food sources, and many other things.

© Eduardo Padilla-Diaz - All Rights Reserved. epadilladiaz@gmail.com

BioBots

🕭 *"These wonderful beings inhabit our body from the molecular to the multi-organic rings."*

8.1 Introduction

- ◆ Reflections on *biobots* (biological robots):
- 🕭◆ A *biobot* is a biological robot that performs specialized and necessary work somewhere in our body.
- 🕭◆ Here, we will stop to explore partially the wonderful world of *biobots*. We will see that our body is made up of trillions of them, and we will encounter the presence of their advanced minds. We will meet some molecular *biobots*, like the kinesin, which walk on tracks, taking tiny steps while pulling their loads, and the dynein *biobots*, which move hanging from branches, like apes do. We will also learn about several types of cellular *biobots*, such as the famous spermatozoa, which move swimming propelled by their motorized tails and are guided by their sense of smell to reach their destination, and the macrophage *biobots* that chase their victims through obstacles, until they catch up and devour them. We will meet organic *biobots* like the heart, multiorgan *biobots* like the blood network, and many others. Ultimately, we will be surprised by the type of work they do and how they do it; however, the most important thing is that we will partially realize what our bodies and minds are made of.

- 🔆 In this incursion, we will only consider the *biobots* of the following minds: molecular, cellular, organelle, organic, organic systems, human operational, and human functional.
  1. 🔆 In the molecular ring, there are more than 1,300 different types of molecular *biobots*, which perform countless functions, including rotary motors, energy generators, cargo transport, protein synthesis, catalysts, muscle contraction, extraction of schematics, component synthesis, waste recycling, motility, and more.
     a. 🔆 Rotary motors are molecular *biobots* that generate energy, propel bacteria and sperm, and more.
        - 🔆 The figure video on the right shows the incredible flagellum motor, similar to the one used by sperm, which drives a tail-whip 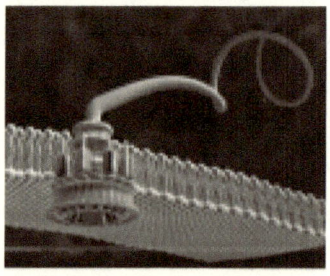 to propel them through fluids. This little machine includes a rotor, stator, drive shaft, U-joint, bushings, bearings, and a whip-like tail that works like a propeller. The machine rotates at 100,000 rpm in one direction. It can reverse direction in a quarter turn and rotate at 100,000 rpm in the other direction.
          - [Amazing Flagellum] Michael Behe : youtube
     b. 🔆 Energy generators: The ATP-Synthase *biobot* is used to generate ATP, the most abundant and most important energy source at the cellular level, and is used by all other types of molecular motors. It has a

rotary molecular motor that uses the flow of protons as a source of energy.
- [ATP synthase] [ATP] [molecular motor] wikipedia

- ⊕ The figure video shows the ATP-Synthase *biobot* in action:
  - [Electron Transport Chain] youtube

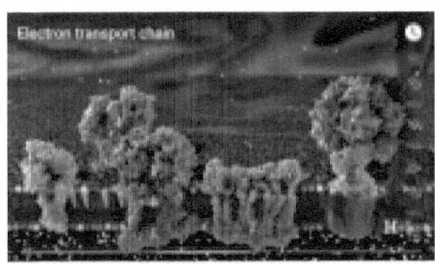

- ⊕ The ATP-Synthase *biobots* are densely packed in the inner mitochondrial membrane, making its entire surface a giant cellular power plant.

  - *The Nobel Prize in Chemistry - 1997 was divided, half to Paul D. Boyer and John E. Walker for their "elucidation of the enzymatic mechanism underlying the synthesis of adenosine triphosphate (ATP)"; and the other half to Jens C. Skou, for the first "discovery of an ion transport enzyme, Na+, K+-ATPase.*
    - [1997 Nobel Prize in Chemistry] nobelprize.org

c. ⊕ Cargo transport: Kinesin, dynein, and myosin are molecular *biobots* that use intracellular networks to transport cargo between different parts of the cell.

- ⊕ Kinesin *biobots* move cargo from the nucleus to the cell periphery. In a cell, there can be up to 45 different types of the kinesin class; Kinesin I moves at 100 steps per second and 8 nanometers per step.
  - [kinesin] wikipedia

  - ⊕ The figure video shows a kinesin walking on a wide track,

- pulling a vesicle through a flexible coupling; the load is thousands of times heavier than him.
  - [Kinesin protein walking on microtubule] youtube
- ⊕ The figure shows the structure of a Kinesin-I. It mainly comprises 1. engine, 2. link, and 3. load coupling.
  - ⊕ The Kinesin-I is implemented by an enzyme, which harnesses ATP energy and converts it into movement or mechanical work. The kinesin is the smallest; the force required to stop it is ~5.4 pN; it is 10 nm long and contains only 340 amino acids; it is about one-third the weight of a myosin and one-tenth the weight of a dynein; it is capable of motility in vitro.
    - [The force generated by a single kinesin ...] pnas.org
- ⊕ Dynein *biobots* move cargo, in retrograde transport, from the periphery to the cell nucleus. These *biobots* also produce the axonemal beating of cilia and flagella, as we will see later. There are 15 different types of dynein-class robots.
  - ⊕ The following figure video shows a dynein *biobot* hanging and swinging to move forward.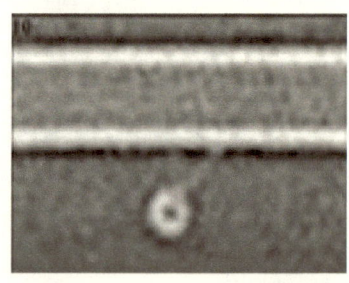

    University of Leeds: "In the laboratory, we started with dynein molecules running along molecular tracks; then, they were frozen in half a step, cooling them at a million degrees per second. Then, using a cryo-electron microscope, thousands of images of the frozen motors were taken in action, combining many

images. Finally, the details of the molecule could be sharpened, and a dynamic idea of its movement could be built."
- [Motor proteins caught 'swinging on monkey bars'] youtube : university of leeds

- ⦿ Myosin *biobots* are responsible for muscle contraction, intracellular cargo transport, and the production of cellular tension. There are 40 different types of the myosin class.

d. ⦿ Protein synthesis: ribosome *biobots* are molecular robots found inside all living cells and carry out the synthesis of biological proteins based on schematics extracted from the genetic knowledge stored in the DNA of the cell nucleus.
- [ribosome] wikipedia

- ⦿ The figure, made by Mariana Ruiz Villareal (ladyofhats : wikimedia), illustrates mRNA translation and protein synthesis by a ribosome.
- [translation (biology)] wikipedia
  [File:Ribosome_mRNA_translation_en.svg] wikimedia

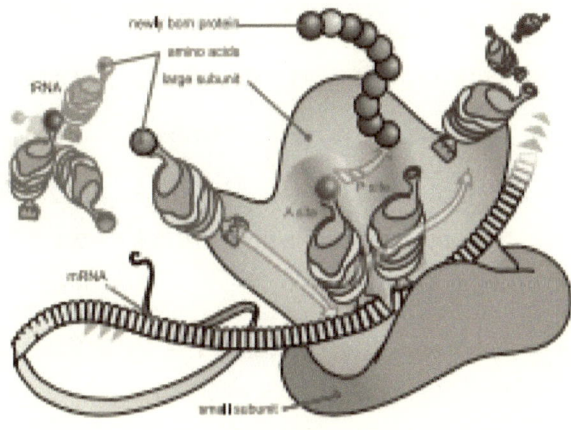

Fig. 8.1

- *The Nobel Prize in Chemistry - 2009 awarded Venkatraman Ramakrishnan, Thomas A. Steitz, and Ada E. Yonath for their studies on "the structure and function of the ribosome."*
  - [1997 Nobel Prize in Chemistry] nobelprize.org

  e. ⊕ Catalysts: The vast majority of molecular robots are chemical robots that fulfill the function of accelerating chemical reactions (catalysts). To put this in perspective, a reaction with orotic acid that would take 78 million years takes only 18 milliseconds with the help of a molecular robot skilled in catalysis, such as OMP-decarboxylase.
     - [biosynthesis] [metabolism] [OMP-decarboxylase] wikipedia

2. ⊕ In the organelle ring, inside each cell, several biobots are organelles. They are its small organs that perform specific jobs, such as:

   a. ⊕ The nucleus (6um) is an organelle-*biobot* which fulfills the following functions:
      - ⊕ It stores the DNA containing blueprints to make all the body parts and proteins.
      - ⊕ It produces the blueprints to make proteins.
      - ⊕ It produces ribosome molecular robots.
      - ⊕ And many more functions.

   b. ⊕ The Golgi apparatus (2 um) is an organelle-*biobot* which functions as a post office and performs the following functions:
      - ⊕ Receives the proteins manufactured by ribosome molecular robots.
      - ⊕ It packages proteins in protective envelopes (vesicles).

- ◈ Mark each envelope with a shipping label (made from special sugar molecules).
- ◈ It sends each envelope (vesicle) with the proteins to one of four possible destinations: to the cell plasma (cytosol), to the cell membrane, to the outside of the cell, or to the waste center (lysosome).

c. ◈ Perizomes (0.1-0.2um: ~100/cell) are organelles-*biobots* responsible for destroying unwanted material.

d. ◈ Lysosomes (0.5um) are organelles-*biobots* that constitute the cell's recycling center.

e. ◈ Mitochondria (1um) are organelles-*biobots* that produce the ATP necessary for the entire cell to function.

f. ◈ The rough plasma reticulum is an organelle-biobot, which, with the help of ribosomes and RNA-schematics, produces proteins, marks them, and transports them to the Golgi apparatus, which is like a mail post office that sends them to their final destination.

g. ◈ The smooth plasma reticulum is an organelle-*biobot* responsible for detoxifying the cell. It produces lipids and steroids, essential in energy storage, membrane structure, and communication (steroids can act like hormones).

3. ◈ In the cellular ring, we find cellular *biobots* since each cell in itself is a biobot; this is how we can find more than 200 types of these in the body, for example, white blood cells, red blood cells, and platelets:

a. ⊕ White blood cells (12-20um) are cellular *biobots* that constitute one of the most essential parts of the immune system. They participate in protecting the body against infectious diseases and foreign invaders. They are found throughout the body, including the blood and lymphatic system.
    - [white blood cells] [pathogens] wikipedia

    - ⊕ The figure video shows a white blood cell of the neutrophil type hunting for a pathogenic agent.

    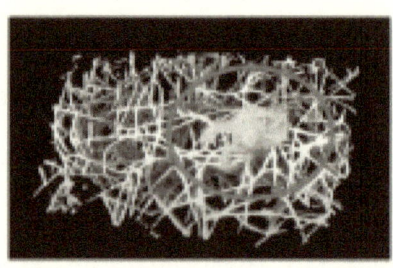

        - [See White Blood Cells Move in 3-D Imaging] youtube

b. ⊕ Red blood cells (6-8um) are cellular *biobots* circulating the blood network continuously. They are like tanks responsible for transporting oxygen from the lungs to the body tissues and eliminating CO_2 in the opposite direction. They are elastic; they narrow in the capillaries to deliver their cargo.
    - [red blood cells] wikipedia

c. ⊕ Platelets (2-3um) are cellular *biobots* that stop bleeding and repair blood vessels. They are found in the blood network. When a blood vessel ruptures, a substance that attracts platelets is exposed from the tissue, and they quickly move to the injury site and adhere to the site in an orderly manner.
    - [platelets] [platelets] [fibrin] wikipedia

    - Platelets change their external shape and join together, forming a layer; they expel ligands into the bloodstream to attract more platelets to create a temporary plug. They then combine with fibrin and other coagulant proteins that circulate in the blood. A clot of fibrin threads forms on the platelet plug, sufficient to stop the bleeding. When it is no longer needed, the clot is removed.

- ⊕ The figure video shows how platelets repair a wound.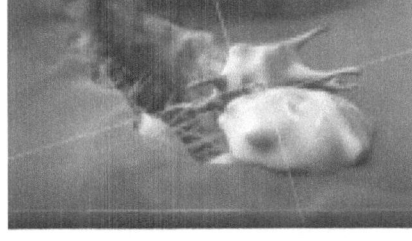
    - [how platelets repair a wound.]
    thrombosis adviser : youtube

4. ⊕ The organic ring has about 100 types of organs/*biobots*, including the heart, brain, lungs, liver, kidneys, bladder, stomach, intestines, and many more.

    - ⊕ Here, we are not going to go into detail about the specific functionality of each of them. What interests us in this context is to know that each organ is a powerful organic robot that performs highly complex motor functions, such as, for example, the heart:
        - [organic systems] [organs of the human body] wikipedia

        - The heart circulates blood through the blood system; it controls valves (see gif figure) to draw oxygenated blood from the lungs, then pours it into the 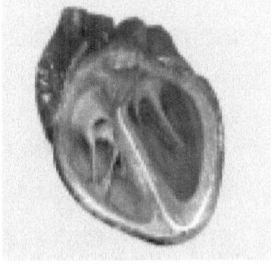 bloodstream and pumps it to all the body's organs, which absorb it through capillaries. It controls valves to suck the deoxygenated blood that the organs pour into the bloodstream to send it back to the lungs to be oxygenated again.
            - [heart] wikipedia

- ⊕ The capillaries in skeletal muscle have valves controlled by capillary *biobots* that close and by capillary *biobots* that close and open them in groups, controlling blood flow through them. They have been the center of study for more than a hundred years. A Nobel Prize has been awarded for discoveries in the way they operate. Still, the algorithm that controls them has not been possible to characterize to date. As a curious fact, the total number of capillaries in adult human beings is such that if all the capillaries formed a continuous tube, they could reach at least twice the earth's circumference.
  - [1920 August Krogh, Nobel Lecture] nobelprize.org

  - *The 1920 Nobel Prize in Physiology or Medicine was awarded to Schack August Steenberg Krogh "for his discovery of the regulatory mechanism of the capillary motor.*
    - [1920 Nobel Prize in Physiology or Medicine] nobelprize.org

5. ⊕ In the ring of the organic systems, countless *biobots* work together to perform very useful functions, such as the following systems: nervous, motor, digestive, respiratory, urinary, reproductive, endocrine, blood, lymphatic, sensory, integumentary, and others.

    - ⊕ Here, we are not going to go into detail about the specific functionality of each of them. This exploration interests us because each comprises several organic robots, several types of tissues, and various fluids, forming a large, highly intelligent robot that performs important functions for one or more organic systems.
      - [organic systems] [organs of the human body] wikipedia

6. ⊕ In the ring of the virtual mind, semi-*biobots* appear, which are semi-autonomous automation

biobots that depend on the conscience to initiate and the inconscience to execute.

- 🛈◈ The semi-biobots in this ring get their automation knowledge genetically.

- 🛈◈ After being initiated by the conscience, these semi-*biobots* operate autonomously until they fulfill their automation purpose.

- 🛈◈ In the ring of the virtual mind, these semi-biobots are used for countless automation functions, such as drinking, eating, walking, remembering, and more.

    - 🛈◈ In the Thinking Center, semi-*biobot* automation is responsible for one of the most important mental functions: remembering; it is done by extracting records from episodic knowledge, synthesizing them, and making us perceive them.

7. 🛈◈ In the ring of the operational human, semi-*biobots* that learn appear, they are acquired automation *biobots* that depend on the conscience to learn and the subconscience to execute.

    - 🛈◈ Learning is a conscious function that leads to the automatization of specific motor sequences, which later we only need to invoke consciously and leave to the subconscience to supervise their execution.

    - 🛈◈ After being initiated by the conscience, these semi-*biobots* operate until they fulfill their acquired automation purpose. The conscience can suspend the automation function of these semi-*biobots* at any time it wishes.

- In the ring of the operational human, semi-*biobot* acquired automation is used in several operational functions, such as more sophisticated drinking, eating, walking, and more.
  - The Vocal Center is full of semi-*biobot* acquired automation; it is behind speech and singing by synthesizing the sound of each word or sequence of words or modulating the sound of words with melodies.
  - The Visual Center is full of semi-*biobot* acquired automation; it is behind reading. This subject is covered in the next book of the series.
8. In the ring of the functional human and all the superior rings, semi-*biobots* are used in countless acquired automation functions, but they are not covered in this exploration.

© Eduardo Padilla-Diaz - All Rights Reserved. epadilladiaz@gmail.com

Molecular Nets

🕭 *"The intracellular networks where molecular robots operate."*

9.1 Introduction

- ◉ Reflections on *Networks* 1 (part 1 of 3):
- ◉ Intracellular *Networks* are used by molecular robots to provide their services, which vary from cargo transportation to muscle contraction. They are infrastructure *Networks* formed by tracks made of proteins, which extend throughout the entire cell, forming part of its skeleton. Here, we will explore the *Wide Tracks* and the *Thin Tracks*.

---

1. ◉ *Wide Tracks*: They are formed of microtubules about 25 nanometers in diameter made of tubulin; they can reach 50 micrometers long. They begin within the nucleus and extend toward the periphery.
    - [microtubules] [tubulin] wikipedia

Fig. 9.1

- ◉ *Wide Tracks* are used in many ways; for example, they are used by kinesin and dynein molecular robots for intracellular cargo transport of mitochondria and vesicles.
    - [kinesin] [dynein] wikipedia

- ⊕ The figure video shows a kinesin robot walking on a *Wide Track*, pulling through a flexible coupling, a load thousands of times heavier than him.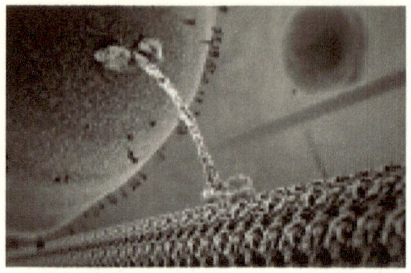
    - [kinesin walking] youtube

2. ⊕ *Thin Tracks*: They are formed of protein microfilaments of about 7-9 nm in diameter, made by two strands of actin.
    - [microfilaments] [actin] wikipedia

$\updownarrow$ 7-9 nm

Fig. 9.2

- ⊕ *Thin Tracks* are used in many ways, including but not limited to cytokinesis in cell division and motility, giving specific cells the ability to move independently in muscle cells; myosin molecular robots use them as traction platforms in muscle contraction.
    - [cytokinesis] [motility] [myosin] wikipedia

© Eduardo Padilla-Diaz - All Rights Reserved. epadilladiaz@gmail.com

Neural Nets

"Neurons connect reality, virtuality, and modulators with body and mind so we can be."

10.1 Introduction

- Reflections on networks (part 2 of 3):
- The *Neural Nets* are a set of intelligent, adaptive, and informatic networks responsible for the operativity and functionality of the body and mind. They are made mainly of neuron cells assisted by glial cells.
  - The *Neural Nets* are explored in 3 parts: Book 2 (The Width of the Present) explores them at the cellular level, Book 3 (Tokens and Words) explores them at the *organic* level, and Book 4 (The Domains of the Conscience) explores them at the virtual and external levels.
- Here, we explore the *Neural Nets* at the cellular level. As stated previously, it is mainly made by different types of neurons, assisted by glial cells:
  - [neurons] [glia cells] wikipedia

1. Neurons interconnect with other neurons and cells, forming highly specialized networks for transmitting signals and information. They have diameters from 5 to 150 μm, which is why they are among the largest and smallest cells. It is estimated that there are 86 billion neurons in the human body.
   - [neurons] wikipedia

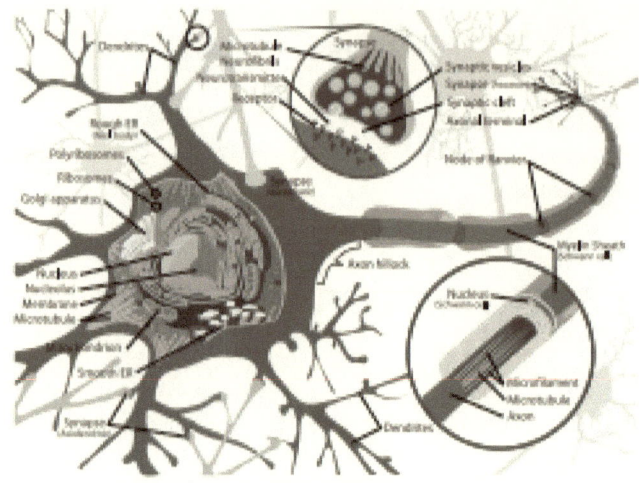

Fig. 10.1

- ◉ The figure, made by Mariana Ruiz Villareal (ladyofhats : wikimedia), shows a typical neuron: it consists of a cell body (soma), input connectors (dendrites), and a single axon where the output connectors (synapses) are located.
    - [neurons] wikipedia
      [File:Complete_neuron_cell_diagram_en.svg] wikimedia
- ◉ Dendrites are the neuron's input connectors, receiving information from other neurons.
    - [dendrite] wikipedia

    - ◉ Dendritic branching can be extensive in a single neuron to receive up to 100,000 inputs.
- ◉ Synapses are the output connectors. They allow neurons to transmit signals and information to other neurons or effector cells.

    - ◉ When we learn something, new neurons are not created; the ones we already have are used, but new synapses are formed. If the synapses are not used, they are erased, but if they are used, they are retained and strengthened.

- ◆ The number of synapses in a typical neuron varies between 1,000 to 10,000.
  - [Brain Facts and Figures] faculty.washington.edu
- ◆ The number of synapses in the cortex is around 150 trillion.
  - [Brain Facts and Figures] faculty.washington.edu
- ◆ The number of synapses in a Purkinje neuron (located in the cerebellum) is around 200,000.
  - [purkinje neuron] wikipedia
- ◆ The figure illustrates this type, hand-drawn by the Spaniard neuroscientist, pathologist, and histologist Santiago Ramon y Cajal, known as the father of modern neuroscience, Nobel 1906.
  - [Santiago Ramón y Cajal] wikipedia

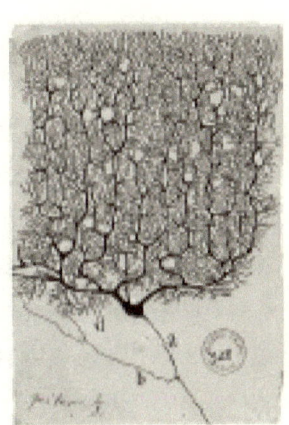

Fig. 10.2

- *The 1906 Nobel Prize in Physiology or Medicine was awarded jointly to Camillo Golgi and Santiago Ramón y Cajal "in recognition of their work on the structure of the nervous system."*
  - [1906 Nobel Prize in Physiology or Medicine] nobelprize.org

- ◆ The following figure graphically shows the different types of synaptic connectors:
  - [Synapse] wikipedia

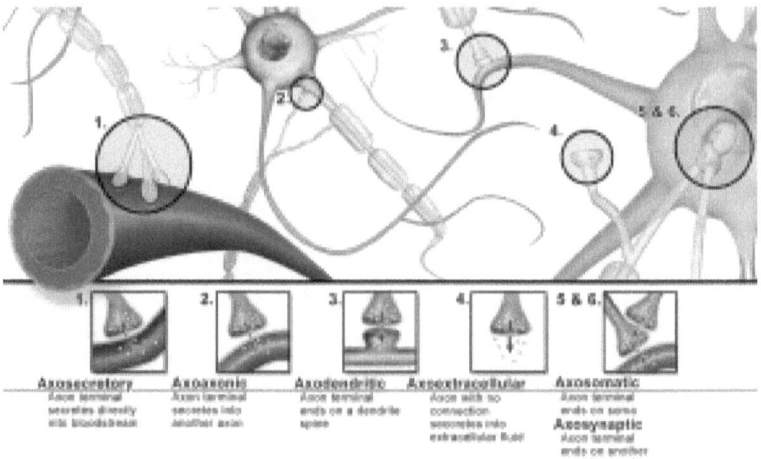

Fig. 10.3

- ◆ Neurons, according to their function, are classified into the following types: sensory, motor, and interneurons:

  - ◆ Sensory neurons (afferents): They couple sensory information to sensory centers. They convert the stimuli from their receptors into useful sensory information sent to different centers, such as the Visual or Motor centers, for processing. The stimulus can come from outside the body, such as light or sound, or inside the body, such as blood pressure or the sense of body position.
    - [sensory neurons] [sensory receptors] wikipedia

    - ◆ The figure shows the different types of sensory neurons, which have different sensory receptors that respond to different stimuli.

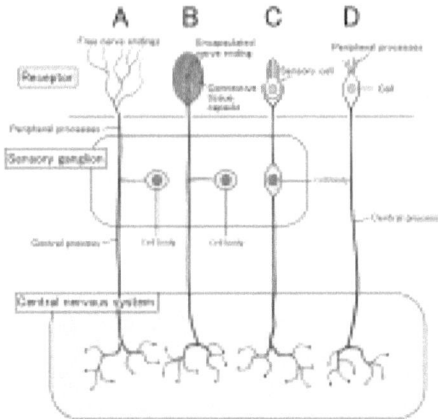

Fig. 10.4

- ◈ Motor neurons: couple sensory centers with muscles and glands; conduct motor commands from the cortex to the spinal cord or from the spinal cord to the muscles; they directly or indirectly control effector organs such as muscles and glands. There are two types of motor neurons: upper and lower:
  - [motor neurons] wikipedia

  - ◈ Upper: Their axons synapse on the interneurons of the spinal cord and occasionally directly on the lower motor neurons.
    - [upper motor neuron] wikipedia

  - ◈ Lower: Their axons are efferent nerve fibers that carry signals and information from the spinal cord to the effectors.
    - [lower motor neuron] wikipedia

- ◈ Interneurons: Couple the flow of signals and information between a sensory neuron or a motor neuron.
  - [interneurons] wikipedia

2. ⊕ Glial cells: Also called glia or neuroglia, they are non-neuronal cells; they maintain constant stability in the internal, physical, and chemical conditions of the nervous system (homeostasis). They are responsible for repairing and regenerating injuries to the nervous system. They serve as insulators in excitable tissues, forming the myelin sheaths that protect and insulate the axons of neurons. They surround neurons and hold them in place; they supply nutrients and oxygen to neurons; they isolate one neuron from another, destroy pathogens and eliminate dead neurons; and much more.
    - [glia] [homeostasis] [myelin] wikipedia

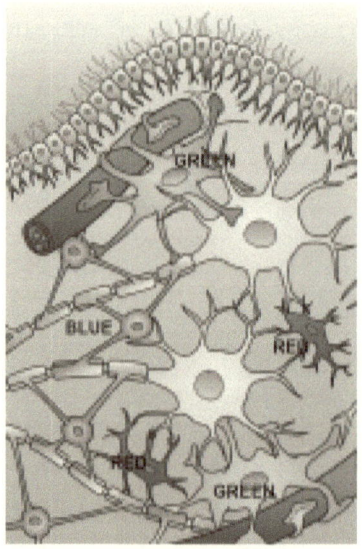

Fig. 10.5

- ⊕ The above figure shows the different types of glial cells found in the central nervous system: astrocytes (green), microglial cells (dark red), and oligodendrocytes (light blue).
    - [types of glia cells] wikipedia

- ◎ The human body is estimated to have 86 billion glial cells.

3. ◎ The following video by the Hoogenraad Lab at Utrecht University explains in detail the fundamental mechanisms of neuronal trafficking, helping us understand how intracellular transport underlies the development and function of nerve cells.
   - [casper hoogenraad] wikipedia [neural trafficking mechanisms] Hoogenraad Lab : youtube

4. ◎ The following video by scientists at the Salk Institute shows how they computationally reconstructed brain tissue in the hippocampus to study the sizes of the connections (synapses).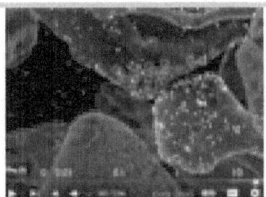
   - [how synapses work] youtube [Salk Institute] wikipedia

5. ◎ The following video (courtesy of Stephen Smith—Stanford Medicine) visualizes the brain as a universe of synapses based on a visual reconstruction (from an array of tomography data) of the synapses in a mouse's somatosensory cortex.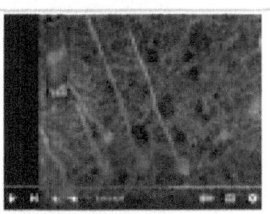
   - [Visualizing the brain as a universe of synapses] youtube

© Eduardo Padilla-Diaz - All Rights Reserved. epadilladiaz@gmail.com

Eduardo Padilla-Diaz

Bot Nets

① *"Networks where cellular robots circulate to transport supplies, repair, and protect."*

11.1 Introduction

- ⊕ Reflections on Networks (part 3 of 4):
- ⊕ Here, we are going to explore the following networks: *Blood Network*, *Lymphatic Network*, and *Endocrine Network*:

1. ⊕ The *Blood Network* is the operational network of the circulatory system. It is a blood transport and distribution system composed of a vascular network, the heart, which circulates blood, and the lungs, which inhale air to extract oxygen and exhale CO_2.

    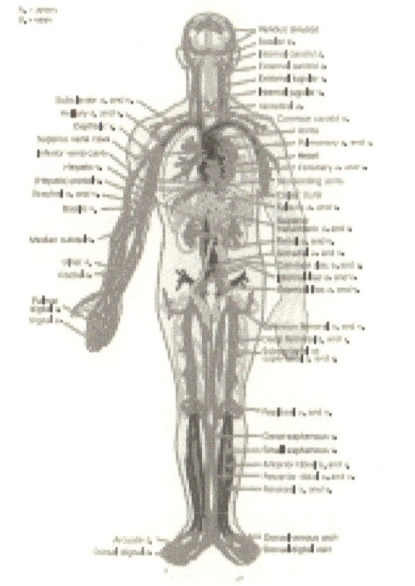

    - [circulatory system] wikipedia
        - ⊕ The figure, made by Mariana Ruiz Villareal (ladyofhats : wikimedia), is an illustration of the *Blood Network*.
            - [File:Circulatory_System_en.svg] wikimedia

a. ◆ The vascular network comprises the arterial subnet, the venous subnet, and the capillaries.
   - ◆ The arterial subnet transports oxygen and nutrients in the blood to the different organs. It begins in the lungs and ends in the capillaries.
   - ◆ The venous subnet transports CO_2, which is discarded by the different organs in the blood; it begins in the capillaries and ends in the lungs.
   - ◆ Capillaries are the smallest blood vessels in the body; they transport blood between arterioles and venules. They are the site of exchange of many substances from the blood with the interstitial fluid surrounding them. Substances that pass through capillaries include water, oxygen, CO_2, and others.
     - [capillary vessels] wikimedia

b. ◆ The figure on the right shows how veins and arteries are found in capillaries.
   - [capillary vessels] wikipedia [File:Blood_vessels-en.svg] wikimedia

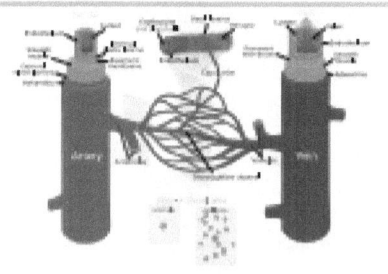

c. ◆ The figure on the right shows how veins and arteries reach bone cells.
   - [cortical bone] wikipedia [File:Illu_compact_spongy_bone.jpg] wikimedia

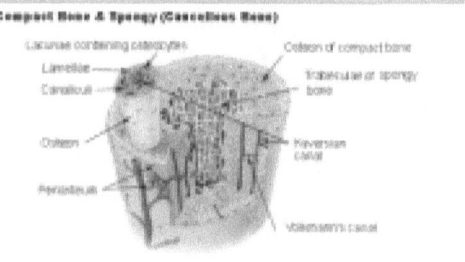

d. ⊕ The figure on the right shows how veins and arteries reach muscle cells.
   - [muscle physiology] doctorlib.info

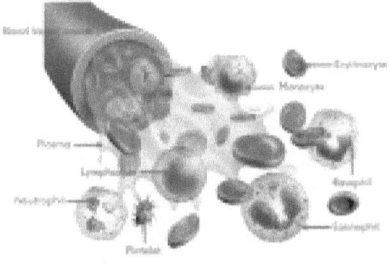

e. ⊕ Blood is made up of plasma, red blood cells, white blood cells, platelets, and other materials:

   - ⊕ The figure on the right shows an illustration of blood with its different components:
     - [blood] physio-pedia.com

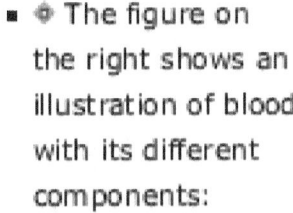

   - ⊕ Plasma (55%): is composed mainly of water (95%). The liquid part of the blood serves as a medium to transport various types of cellular robots and other substances through the *Blood Network*.
     - [plasma] wikipedia

   - ⊕ Red blood cells (44%): They are cellular robots that are tankers used to transport oxygen from the lungs to all the cells in the body and to eliminate CO_2 in the opposite direction.
     - [red blood cells] wikipedia

   - ⊕ White blood cells (<1%) are cellular robots specialized in pursuing and hunting pathogenic agents - such as viruses and bacteria that harm the body.
     - [white blood cells] [pathogens] wikipedia

   - ⊕ Platelets (<1%) are cellular robots that repair blood vessels and stop bleeding.
     - [platelets] wikipedia

- ⊕ And other materials such as proteins, electrolytes, CO2, and hormones.

2. ⊕ The *Lymphatic Network* shown on the right, forms the third line of defense of the human body; it is mainly involved with the immune response. It is where the grand army, composed of around 2 trillion lymphocytes, resides and operates. These lymphocytes are cellular robots that combat pathogenic agents that can harm the body. These robots are made to evolve through an epigenetic mechanism, which causes them to specialize as they are exposed to combat.

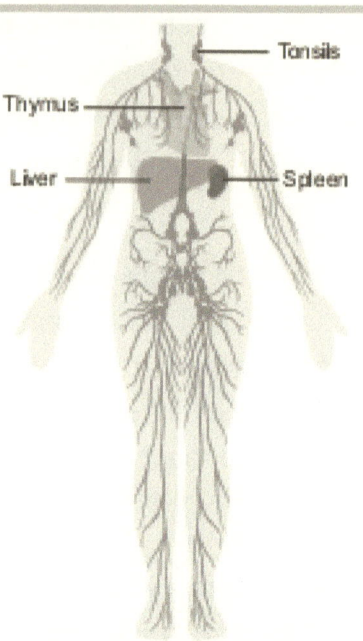

 - [Diagram_of_the_lymphatic_system_CRUK_041.svg] wikimedia commons [number of lymphocytes] harvard.edu [lymphatic system] [epigenetic] wikipedia

- ⊕ The immune response fights specific pathogens, allowing the immune system to "remember" the pathogen after the infection. Suppose the pathogen tries to invade the body again. In that case, the immune response against that pathogen is much faster and more robust.
    - [lymphatic system] nigerian scholars

- Nobel Prizes and the Immune System:

  1. ⊕ *The 1901 Nobel Prize in Physiology or Medicine was awarded to Emil Adolf von Behring "for his work on serum therapy, especially its application against diphtheria, with which he has opened a new path in the domain of medical science and thus put into practice "in the hands of the doctor a victorious weapon against illness and death."*
     - [1901 Nobel Prize in Physiology or Medicine] nobelprize.org

  2. ⊕ *The 1908 Nobel Prize in Physiology or Medicine was awarded jointly to Ilya Ilyich Mechnikov and Paul Ehrlich "in recognition of their work on immunity."*
     - [1908 Nobel Prize in Physiology or Medicine] nobelprize.org

  3. ⊕ *The 1913 Nobel Prize in Physiology or Medicine was awarded to Charles Robert Richet "in recognition of his work on anaphylaxis."*
     - [1913 Nobel Prize in Physiology or Medicine] nobelprize.org

  4. ⊕ *The 1919 Nobel Prize in Physiology or Medicine was awarded to Jules Bordet "for his discoveries related to immunity."*
     - [1919 Nobel Prize in Physiology or Medicine] nobelprize.org

  5. ⊕ *The 1930 Nobel Prize in Physiology or Medicine was awarded to Karl Landsteiner "for his discovery of human blood groups."*
     - [1930 Nobel Prize in Physiology or Medicine] nobelprize.org

  6. ⊕ *The 1960 Nobel Prize in Physiology or Medicine was awarded jointly to Sir Frank Macfarlane Burnet and Peter Brian Medawar "for discovering acquired immunological tolerance."*
     - [1960 Nobel Prize in Physiology or Medicine] nobelprize.org

  7. ⊕ *The 1972 Nobel Prize in Physiology or Medicine was awarded jointly to Gerald M. Edelman and Rodney R. Porter "for their discoveries on the chemical structure of antibodies."*
     - [1972 Nobel Prize in Physiology or Medicine] nobelprize.org

8. ⊕ *The 1980 Nobel Prize in Physiology or Medicine was awarded jointly to Baruj Benacerraf, Jean Dausset, and George D. Snell "for their discoveries on genetically determined structures on the cell surface that regulate immunological reactions."*
   - [1980 Nobel Prize in Physiology or Medicine] nobelprize.org

9. ⊕ *The Nobel Prize in Physiology or Medicine 1984 was awarded jointly to Niels K. Jerne, Georges JF Köhler, and César Milstein "for theories on specificity in the development and control of the immune system and the discovery of the principle for the production of monoclonal antibodies."*
   - [1984 Nobel Prize in Physiology or Medicine] nobelprize.org

10. ⊕ *The 1987 Nobel Prize in Physiology or Medicine was awarded to Susumu Tonegawa "for discovering the genetic principle for generating antibody diversity.*
    - [1987 Nobel Prize in Physiology or Medicine] nobelprize.org

11. ⊕ *The 1996 Nobel Prize in Physiology or Medicine was awarded jointly to Peter C. Doherty and Rolf M. Zinkernagel "for their discoveries on the specificity of cell-mediated immune defense.*
    - [1996 Nobel Prize in Physiology or Medicine] nobelprize.org

⊕ The *Lymphatic Network* is the operational network of the lymphatic system and part of the immune system. It extends throughout the body and comprises lymph, lymphatic vessels, lymphocytes, lymph nodes, lymphatic valves, lymphatic capillaries, lymphoid organs, and part of the *Blood Network*.
- [lymphatic system] [immune system] wikipedia

a. ⊕ Lymph: is a liquid similar to blood plasma; It serves mainly as a carrier medium for cellular combat robots, lymphocytes, and other materials.
   - [lymph] wikipedia

b. ◉ Lymphatic vessels make up the network, which extends throughout all body tissues. They transport lymph from the lower part of the body to the heart and drain from the head, shoulders, and arms. Lymphatic vessels help the circulatory system and all the cells in the body by removing waste, germs, and excess water from tissue fluid.
    - [lymphatic vessels] wikipedia

c. ◉ Lymphocytes are cellular robots specialized in combat and are part of the immune system; they are white blood cells. There are two main types: T-cells and B-cells:
    - [lymphocytes] wikipedia

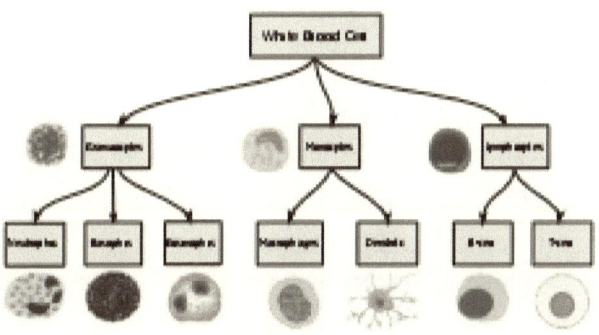
Fig. 11.1

- ◉ The figure shows the taxonomy of bone marrow white blood cells, including three main types: granulocytes, monocytes, and lymphocytes; and seven subtypes: neutrophils, basophils, eosinophils, macrophages, dendritic cells, T-cells and B-cells.
    - [white blood cell taxonomy] researchgate.net
- ◉ T-cells develop in the thymus and combat foreign invaders directly. They produce cytokines that activate other cellular fighting robots called macrophages, which act to clear invaders and dead

tissue after an immune response. The T-cells circulate the Lymphatic Network, enter and circulate the Blood Network, then circulate the tissues, re-enter the *Lymphatic Network*, and repeat the cycle.
- [T-cells] [cytokines] [macrófagos] wikipedia

    - ◈ Naive T-cells can recognize antigens from virus-infected cells or tumor cells and, in response to those antigens, will undergo vigorous clonal expansion where even a single naïve T-cell can become hundreds of thousands of activated effector cells that go on to be responsible for killing cells infected by pathogens or tumor cells. Most of these effector T-cells will die after the infection has cleared or the tumors have been controlled. Only a stable population of memory T-cells will persist in the coming weeks, and they can respond again and become a population of secondary effector T-cells that can combat the infected cell or the tumor cell.
        - [T cell types and their functions explained by Erika Pearce] youtube

- ◈ B-cells develop in the bone marrow, and through differentiation, they evolve into more specialized forms when they have been exposed to an antigen:
    - [B-cells] [differentiation] wikipedia

    - ◈ A naive B-cell has not been exposed to an antigen. Once exposed, it becomes a B-cell with memory (memory B-cell) or a plasma B-cell:
        - ◈ Memory B-cells are formed within the germinal centers after primary infection; they can survive for decades and repeatedly generate an antibody-mediated secondary immune

response, which is accelerated and robust in the case of reinfection.
- [memory B-cells] wikipedia

- ● The plasma-B-cells originate in the bone marrow and secrete large amounts of antibodies in response to antigen presentation. These antibodies are transported from the plasma-B-cells by the blood plasma and lymphatic system to the target antigen (foreign substance) site, where they neutralize or destroy. B-cells differentiate into plasma-B-cells that produce antibody molecules closely patterned after the receptors on precursor B-cells.
  - [plasma B-cells] wikipedia

d. ● The lymphatic organs: The figure illustrates the lymphatic organs and their relationship with the Lymphatic Network. They are divided into two groups: primary and secondary:
- [lymphatic organs] wikipedia
  [File:2201_Anatomy_of_the_Lymphatic_System.jpg] wikimedia

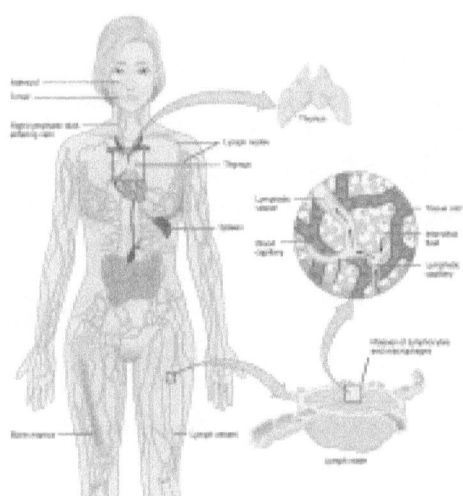

Fig. 11.2

- ◉ Primary: They are where lymphocytes develop and are the bone marrow and the thymus:
    - ◉ Bone marrow: Found within many bones; B-lymphocytes form here.
        - [bone marrow] wikipedia
    - ◉ The thymus in the upper part of the chest behind the breastbone stores and develops T-lymphocytes.
        - [timo] wikipedia
- ◉ Secondary: After the lymphocytes are formed, they migrate to these organs where they can react with a foreign antigen, and they are the spleen, the tonsils, and the lymph nodes:
    - ◉ The spleen is in the upper part of the abdomen. It filters pathogens and worn-out red blood cells from the blood, and then lymphocytes in the spleen destroy them.
        - [spleen] wikipedia
    - ◉ Tonsils are located in the throat on both sides of the pharynx. They trap pathogens, which are destroyed by lymphocytes in the tonsils.
    - ◉ The lymph nodes, around 600 in total, are strategically distributed in the network. As lymph flows through them, they control and filter the composition of bacteria, cancer cells, and other life-threatening agents.
        - [lymph nodes] wikipedia

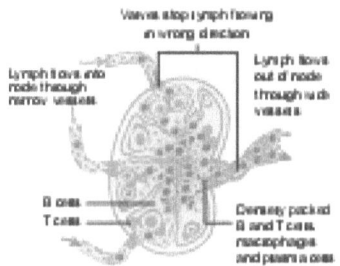

Fig. 11.3

- ⊕ The figure shows a lymph node filled with cellular combat robots, such as T-lymphocytes, B-lymphocytes, macrophages, and plasma cells.
  - [File:Diagram_of_a_lymph_node_CRUK_022.svg] wikimedia

e. ⊕ Lymphatic capillaries are vessels where interstitial fluid enters the lymphatic system to become lymphatic fluid. They are located in almost all tissues of the body.

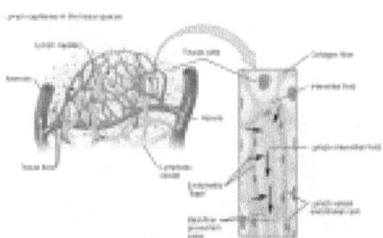

Fig. 11.4

- ⊕ The figure shows how lymphatic capillaries intertwine between the arterioles and venules of the circulatory system in the soft connective tissues of the body. The exceptions are the central nervous system, bone marrow, bones, teeth, and cornea of the eye, which do not contain lymphatic vessels.
  [Lymphatic System] wisc.pb.unizin.org
  [File:Diagram_of_a_lymph_node_CRUK_022.svg] wikimedia

f. ◉ The circulatory system includes the lymphatic system, which circulates lymph. Lymph passage takes much longer than that of blood. After being filtered from the interstitial fluid (between cells), lymph is recycled into excess blood plasma and returned to the lymphatic system.

3. ◉ The *Endocrine Network* is the operational network of the endocrine system. It uses the *Blood Network* to transport the hormones secreted by the endocrine glands to deliver them to specific cells in distant organs and tissues. The endocrine system is made up of a *Blood Network*, endocrine glands, and hormones:
    - [endocrine system] wikipedia [endocrine system] youtube

    a. ◉ The endocrine glands secrete hormones directly into the *Blood Network*.
        - [endocrine glands] wikipedia

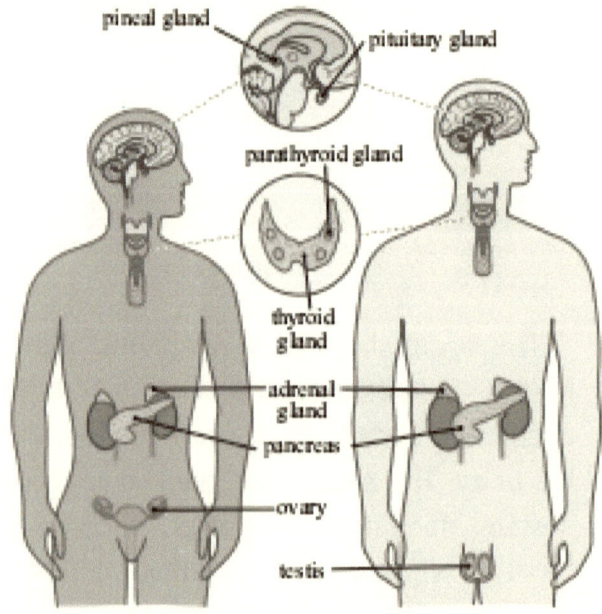

Fig. 11.5

- ◉ The figure shows the location of the different endocrine glands: the pineal gland, the pituitary gland, the pancreas, the ovaries, the testicles, the thyroid gland, the parathyroid gland, the hypothalamus, and the adrenal glands. The hypothalamus and pituitary glands are neuroendocrine organs.
    - [File:Endocrine_English.svg] wikimedia

b. ◉ Hormones are a class of signaling molecules that are transported to specific cells of distant organs and tissues to regulate many physiological processes and behavioral activities such as digestion, metabolism, respiration, sensory perception, sleep, excretion, lactation, stress induction, growth and development, movement, reproduction, and mood manipulation.
    - [hormones] wikipedia

© Eduardo Padilla-Díaz - All Rights Reserved. epadilladiaz@gmail.com

Eduardo Padilla-Diaz

Signaling

⚜ *"Signals used in the neural and cellular networks to exert actions in cells."*

12.1 Introduction

- ⚜ Reflections on *Cellular Signaling* at the operational level:
- ⚜ *Cellular Signaling* is how cells send signals to other cells to exert actions in them. For example, motor neurons activate muscles by sending them signals.
  - ⚜ Essentially, the *Cellular Signaling* mechanism is made up of three parts: *Senders*, *Receivers*, and *Signals*:
    - ⚜ *Senders* are cells with output *signaling* connectors that allow them to send signals to other cells or themselves.
    - ⚜ *Receptors* are cells with input *signaling* connectors that allow them to receive signals from other cells or themselves.
    - ⚜ *Signals*: They are made up of certain types of molecules that can only be received by cells with *Receptors* for said types.
  - ⚜ Here, we will explore the following types of signaling: *Neural*, *Endocrine*, and *Immune*:
    - [cell signaling] wikipedia

- ◉ In *Neural Signaling*, neurotransmitters are secreted by neurons through synapses to deliver them to other cells in contact, such as another neuron or a muscle cell, to exert a specific action on them.
  - [neurotransmitters] [nervous system] wikipedia

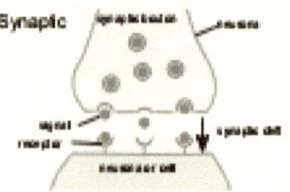

- ◉ In *Endocrine Signaling*, hormones are secreted by endocrine cells and transported through the blood network to distant locations to act on target cells.
  - [hormones] [endocrine system] wikipedia

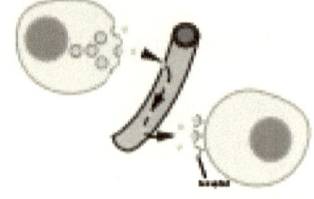

- ◉ In *Immune Signaling*, cytokines are secreted by immune system cells, such as white blood cells. Cytokines diffuse through the lymphatic and blood networks to act on target cells.
  - [cytokines] [immune system] wikipedia

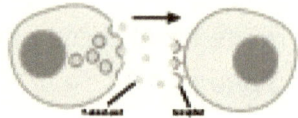

- ◉ *Signals*: They are made up of specific molecules that cells can only receive with *Receptors* for said signals. Here, we will consider the following: *Neurotransmitters*, *Hormones*, and *Cytokines*:

1. ◉ *Neurotransmitters* are the main *signaling* molecules of the nervous system; neurons produce them. The exact types of *Neurotransmitters* unique to humans are unknown, but more than 200 have been identified. Here, we are going to mention the following: *Acetylcholine* and *Dopamine*:

- *Acetylcholine* (amino acid) is the *Neurotransmitter* motor neurons use to activate the muscles. The figure shows its molecular model.
    - [acetylcholine] [molecular model] wikipedia
    - *The 1936 Nobel Prize in Physiology or Medicine was awarded jointly to Sir Henry Hallett Dale and Otto Loewi "for their discoveries relating to the chemical transmission of nerve impulses. Otto Loewi demonstrated the function of acetylcholine as a messenger between nerves and organs."*
        - [1936 Nobel Prize in Physiology or Medicine] nobelprize.org
- *Dopamine* is a *Neurotransmitter* released when we do something that makes us feel good. The figure shows its molecular model.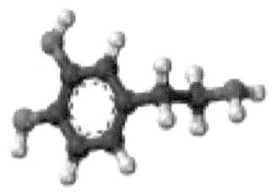
    - [dopamine] [molecular model] wikipedia
    - *The 2000 Nobel Prize in Physiology or Medicine was awarded jointly to Arvid Carlsson (dopamine), Paul Greengard, and Eric R. Kandel "for their discoveries on signal transduction in the nervous system and synapses' central role in memory and learning.*
        - [2000 Nobel Prize in Physiology or Medicine] nobelprize.org

2. *Hormones* are the main *signaling* molecules of the endocrine system. They are produced in endocrine glands, such as the testicles, ovaries, and others. Here, we will mention the following: *Testosterone, Estrogen, Oxytocin, Adrenaline, Insulin, Levothyroxine,* and *Cortisol*.
    - [testicles] [ovaries] wikipedia

- ✿ *Testosterone* is the main male sex hormone. It is produced in the testicles and plays a key role in 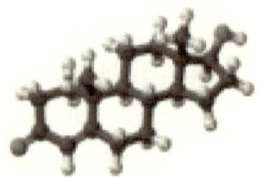 developing them and the prostate. It promotes increased muscle and bone mass and body hair growth, participates in health and well-being, and prevents osteoporosis. The figure shows its molecular model.
  - [testosterone] [molecular model] wikipedia

  - ✿ *The 1939 Nobel Prize in Chemistry was divided equally between Adolf Friedrich Johann Butenandt "for his work on sex hormones (testosterone and estradiol/estrogen)" and Leopold Ruzicka "for his work on polymethylenes and higher terpenes."*
    - [1939 Nobel Prize in Chemistry] nobelprize.org

- ✿ *Estrogen* is the main female sex hormone; it is produced in the ovaries; participates in the regulation 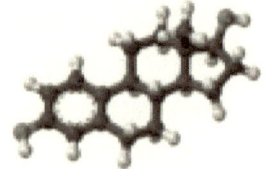 of menstrual cycles; it is responsible for the development of breasts and the widening of hips; it is essential in the development and maintenance of the mammary glands, uterus, and vagina; it also has effects on bones, fat, skin, liver and brain. The figure shows its molecular model.
  - [estrogen] [estradiol] [molecular model] wikipedia

  - ✿ *The 1939 Nobel Prize in Chemistry was divided equally between Adolf Friedrich Johann Butenandt "for his work on sex hormones (testosterone and estradiol/estrogen)" and Leopold Ruzicka "for his work on polymethylenes and higher terpenes."*
    - [1939 Nobel Prize in Chemistry] nobelprize.org

- ✱ *Oxytocin*, the love hormone, is produced in the hypothalamus; it is released into the bloodstream in response to love and childbirth; this helps with birth, bonding with the baby, and milk production.
  - [oxytocin] [hypothalamus] wikipedia

  - ✱ *The 1955 Nobel Prize in Chemistry was awarded to Vincent du Vigneaud "for his work on biochemically important sulfur compounds, especially for the first synthesis of a polypeptide hormone (oxytocin and vasopressin)."*
    - [1955 Nobel Prize in Physics] nobelprize.org

- ✱ *Adrenaline* is produced in the adrenal glands and some neurons in the medulla. Also known as epinephrine, it is a hormone and medication that regulates visceral functions such as breathing. It plays an important role in the 'fight or flight' response by increasing blood flow to the muscles, heart pumping capacity, pupil dilation, and blood sugar level. The figure shows its molecular model.
  - [adrenaline] [adrenal glands] [fight or flight] [molecular model] wikipedia

  - ✱ *The 1971 Nobel Prize in Physiology or Medicine was awarded to Earl W. Sutherland, Jr. "for his discoveries concerning the mechanisms of action of hormones (adrenaline)."*
    - [1971 Nobel Prize in Physiology or Medicine] nobelprize.org

- ✱ *Insulin* is a hormone produced by specific cells in the pancreas. It helps the body convert food into energy and controls blood sugar levels. Insulin therapy is often an essential part of diabetes treatment.

- [insulin]] [anabolic] [diabetes] wikipedia

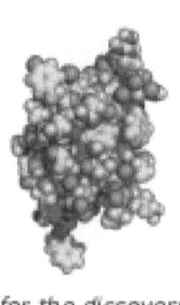

  - ◆ *The 1923 Nobel Prize in Physiology or Medicine was awarded jointly to Frederick Grant Banting and John James Rickard Macleod "for the discovery of insulin."*
    - [1923 Nobel Prize in Physiology or Medicine] nobelprize.org
  - ◆ *The 1958 Nobel Prize in Chemistry was awarded to Frederick Sanger "for his work on the structure of proteins, especially that of insulin."*
    - [1958 Nobel Prize in Chemistry] nobelprize.org

- ◆ *Cortisol* is the primary stress hormone.

  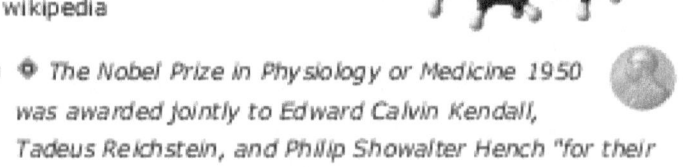

  - [cortisol] [molecular model] wikipedia

  - ◆ *The Nobel Prize in Physiology or Medicine 1950 was awarded jointly to Edward Calvin Kendall, Tadeus Reichstein, and Philip Showalter Hench "for their discoveries relating to the hormones of the adrenal cortex, their structure, and biological effects."*
    - [1950 Nobel Prize in Physiology or Medicine] nobelprize.org

- ◆ *Levothyroxine*, or L-thyroxine, is a synthetic form of the thyroid hormone thyroxine (T4). It is used to treat thyroid hormone deficiency (hypothyroidism). When taken daily, levothyroxine has a half-life of 7.5 days, so about six weeks is required to reach a steady level in the blood.
  - [T4] [molecular model] wikipedia

- *The Nobel Prize in Physiology or Medicine 1909 was awarded to Emil Theodor Kocher "for his work on the physiology, pathology, and surgery of the thyroid gland."*
  - [1909 Nobel Prize in Physiology or Medicine] nobelprize.org

3. *Cytokines* are the main *signaling* molecules of the immune system (Nobels Med 2018, Med 1996, Med 1984). This subject will be explored later in more detail.
   - [cytokines] [[immune system] wikipedia

   - They are essential in the body's immune responses to infections, inflammation, and trauma.

   - They are produced by a wide range of cells, including immune cells such as macrophages, B-cells, T-cells, and more that are not covered in this exploration.
     - [macrophages] [B-cells] [T-cells] wikipedia

© Eduardo Padilla-Diaz - All Rights Reserved. epadilladiaz@gmail.com

Eduardo Padilla-Diaz

Memories-2

"These operational memories store the innate, immediate, and acquired knowledge."

13.1 Introduction

- Reflections on *memories* (part 2 of 3):
- Memory is a device where information is stored to preserve, organize, and use in the future.
  - [memory] wikipedia
- Here, we are going to explore *memories* at an operational level.
- Operational *memories* are classified as *Innate*, *Immediate*, and *Acquired*:

  1. *Innate Memories*: They store the innate knowledge extracted from the genetic code and expressed in various ways, such as innate vocabularies, innate grammar, reflexes, balance system, instincts, and gradual processes. These are explored in more detail next in the Knowledge Chapter.

  2. *Immediate Memories* are modal and multi-context since they exist in independent sets for each issue. They are found in *thinkers*. They are temporary *memories*. They store the immediate knowledge being consciously perceived through sequences of thinking messages, which inform us in the present of reality, virtuality, emotionality, and spirituality. They allow us to analyze and resolve what we will do in the immediate

future and the near future. Thinking Messages are covered later in their chapter.

- *Immediate Memories* are memory sets used to store sequences of thinking messages spread over time, in which the order of each message in the sequence has to be preserved. Let us remember that thinking messages form sequences and that the semantic value of each message is determined by its position in the sequence.

  - As we will see later, the meaning accumulator found in each *thinker* stores its received sequences independently; this allows us to sustain several conversations with several people simultaneously without getting confused.

- At the end of a conscious cycle, the *Immediate Memories* of the present accumulate chronologically and form the immediate past; let's remember that:

  - The immediate past segment is the second most important. In chronological order, this segment contains the most relevant of what was done in the last few present segments. The information in this segment is essential for understanding language, anticipating, protecting ourselves, and many other things.

3. *Acquired Memories* are modal and found in each sensory center. They are intrinsically associative since they are underlain at a physiological level by specialized neurons, which expand their relationships through their synapses. They form associations with very high plasticity.

- 🕐❖ *Acquired Memories* store the acquired knowledge. They are of variable persistence and become as permanent as the neurons and synapses that underlie them. The persistence of each acquired memory is linked to its learning and the frequency with which it is used.

Eduardo Padilla-Diaz

Knowledge

"It is where we store what we know to help us decide what to do next."

14.1 Introduction

- Reflections on Knowledge:
- Knowledge is what we know innately, what we learn through experiences and formal learning, what we know about the reality of the present, and what we know about the future.
  - We need Knowledge to understand, comprehend, reason, and interact with the domains of the conscience to make decisions.
  - Note: Effects that modulate Knowledge, such as emotions and affections, give rise to subjective Knowledge and Knowledge relativity, known as qualia. This aspect will be discussed later.
    - [qualia] wikipedia
- Knowledge is divided into the following branches: Innate, Immediate, and Acquired:

  1. *Innate Knowledge*: It is the one we have by nature; it does not have to be learned; it is unconscious, and it is mainly manifested in all cellular processes, organic processes, and the locomotor system.
     - [locomotor system] wikipedia

- By the time the first mind rings emerged, some form of innate Knowledge was present in each ring. The existence of any ring mind cannot be conceived without some form of innate Knowledge, which is its operational foundation.
- Innate Knowledge: It is extracted from the genetic code. At the human level, it is expressed in various ways, such as innate vocabularies, innate grammar, reflexes, balance, instincts, gradual processes, and other forms like the movement fluidity system, which we will not discuss here.

  a. Innate vocabularies: found in sensory centers. They are modal. They allow the sensory centers to extract objects of interest from the sensory fields in an abstract way.
     - They allow us to abstractly recognize people, animals, plants, and things innately.
     - They are part of the semantic Knowledge.
     - Innate vocabularies guarantee the successful extraction of objects of interest from sensory fields. They are extracted abstractly from the innate vocabularies if they do not exist in the acquired vocabularies.

  b. Innate grammar, better known as universal grammar, was formulated by Noam Chomsky and establishes that the faculty of language has a genetic component and a set of structural rules that are innate and independent of sensory experience.
     - [universal grammar] wikipedia

c. ⊕ Reflexes are automatic responses to stimuli that do not require conscious thinking. They provide us with countless necessary skills for the body to operate optimally and entirely without the help of the conscience. In the body, we have more than 50 reflexes; let's see among them: sneezing, yawning, and innate attention.
    - [reflexes] [list of reflexes] wikipedia

    - ⊕ Sneezing is a reflex that expels mucus containing foreign or irritating particles and cleans the nasal cavity.
        - [sneeze] wikipedia
    - ⊕ Yawn is a reflex that consists of the simultaneous inhalation of air and stretching of the eardrums, followed by exhalation.
        - [yawn] wikipedia
    - ⊕ Innate attention is a modal reflex that occurs in an informatic source (generating sensations) due to specific qualitative attributes that could interest the conscience, such as a sudden change in color, shape, or position.
    - ⊕ Note: Most reflexes are not strictly immutable; they are flexible and can be substantially modified to suit behavioral requirements.
        - [reflex modulation] wikipedia

d. ⊕ The balance system, located in the Cerebellum, maintains balance by adjusting body posture and modulating motor commands to compensate for changes in body position or muscle load. This system is entirely automatic and requires very little or no thought assistance.
    - [Cerebellum] neuroscience online

e. ✦ Instincts are complex patterns of behavior that do not result from a consciously rational process. They are carried out without being based on previous experience and without learning. They must be distinguished from reflexes, which are simple responses to a specific stimulus.
    - [habit and instinct] google books

    - ✦ The following figure video shows a gecko hunting the mouse pointer, causing almost automatic movement in response to a sensory illusion. The pointer acts as a decoy to influence the animal's instinctive behavior. This video won the third prize in the Wiki Science Competition in France in the Fauna and Nature category.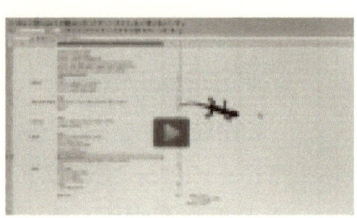
        - [instinct] wikipedia

    - ✦ Maternal instinct: manifested particularly in response to infant crying, it has been respected as one of the most powerful. Functional MRI observations of the mother's brain have partly clarified its mechanism.
        - [instinct] wikipedia

    - ✦ Fear: Experiments on 6-month-old babies demonstrate an increase in pupil dilation when faced with a possible threat from the presence of spiders and snakes.
        - [babies react more excitedly to spiders and snakes] ncbi.nlm.nih.go

f. ✦ Gradual processes are embedded in our genetics; they occur in our bodies unconsciously and manifest themselves some time after our arrival into this world. Here, we will consider the following: puberty and externalization:

- ⚛ Puberty is a process of physical changes through which the body of a boy/girl matures and becomes an adult body capable of sexual reproduction.
  - [puberty] wikipedia

  - ⚛ On average, girls begin puberty at ages 10 to 11 and complete it at ages 15 to 17; boys generally begin puberty at ages 11 and 12 and complete it at ages 16 and 17.
    - [puberty] wikipedia

- ⚛ Externalization: Its primary purpose is to externalize the internal mechanism of the language of the mind (lingua mentis), which is designed to communicate with each other.

2. ⚛ *Immediate Knowledge*: It is what is being consciously perceived through sensory sources that produce sensations in us that inform us about the present, such as reality, virtuality, motor, emotions, affects, and the spiritual, and that allow us to analyze and resolve what we are going to do, in the immediate future and the near future. Immediate Knowledge is the basis of acquired Knowledge.

3. ⚛ *Acquired Knowledge*: It was acquired consciously through immediate Knowledge, experiences, and learning, whether empirical or taught. The detailed classification of this Knowledge would be very extensive and would miss out on the purpose of this exploration. For now, we will only briefly describe the following:

procedural-motor, semantic, declarative, episodic, linguistic, and habitat.

a. ⊕ Procedural-motor Knowledge allows us to consciously improve and automate motor functions executed semi-automatically under the subconscience. The fourth book of the series explores the subconscious.

- ⊕ For example, walking is a motor function that requires countless high-precision motor procedures that we must learn and which are responsible for moving the legs and feet so that we can take steps in different directions, at various speeds, and on diverse terrain.

- ⊕ For example, driving a car is a motor function that requires countless high-precision motor procedures that we must learn and which are responsible for moving the legs and feet to gradually brake, gradually accelerate, maintain a certain speed, move the arms and hands to rotate the steering wheel, keeping the car on the track after changing tracks. All this was learned consciously, but after already learned, it is executed semi-automatically, with minimal conscious intervention.

b. ⊕ Acquired semantic Knowledge describes the Knowledge we have of the words or the cognitive tokens, which represent the objects of interest and mainly their grammatical category. The next series book, Tokens and Words, discusses this in more detail.

- ◆ For example, the word dog is a linguistic signature (sequence of characters) that allows our understanding to find the semantic Knowledge we have of the dog, which is represented by attributes, such as:
    - ◆ It is a noun (grammatical category), is singular (number), is masculine (gender), is an animal, is a mammal, is canine, and many more attributes.
- ◆ The grammatical category of each word or cognitive token makes it easier to understand and extract the meaning of sentences made up of sequences of cognitive tokens or words.
- ◆ The acquired semantic Knowledge is stored in vocabularies at each sensory center.
    - ◆ These vocabularies are modal, which means that the visual center has its visual vocabulary, and the auditory center has its auditory vocabulary.
    - ◆ Vocabularies allow sensory centers to recognize known objects of interest present in their sensory fields.

c. ◆ Declarative Knowledge describes the relationship that objects of interest have with space and matter, how objects of interest are made, and their relationship with other objects of interest.
    - ◆ For example, my dog's name is "chany" and she is very playful.

d. ⊕ Episodic Knowledge: describes the relationship that objects of interest have with time segments and conscient domains.

   - ⊕ For example, my dog's name is "chany" and she is very playful.

e. ⊕ Linguistic Knowledge: allows the exchange of Knowledge between two or more minds based on analogous semantic understanding.

   - ⊕ For example, all students in a class understand what a teacher says because they all have analogous semantic Knowledge.

f. ⊕ Habitat Knowledge describes the environment where we can exist through cognitive maps. The next series book, Tokens and Words, discusses this in more detail.
   - [hippocampus-cognitive maps] wikipedia

   - ⊕ For example, specific routes and points of interest allow us to find shelter and food.

- ⊕ The term Knowledge will be used interchangeably in this book to refer to all types of Knowledge listed in this chapter.

© Eduardo Padilla-Diaz - All Rights Reserved. epadilladiaz@gmail.com

Issues

"Gradual conscient activities processed by thinkers at the thinking center."

15.1 Introduction

- Reflections on *Issues* and *Themes*:
- *Issue*: It is a need, complaint, desire, wish, interest, obligation, goal, or something that is being resolved, that must be resolved, or that has already been resolved.
  - *Issues* are gradual processes controlled by the conscience with the help of attention, processed by *thinkers* in the thinking center.
  - The conscience determines the priority with which *issues* are resolved. The priority of each *issue* can change suddenly according to the results of the last thinking cycle.
- *Theme*: It is a thinking context common to a group of *Issues* that are resolved jointly.
  - For example, paying the phone bill, which is an *issue*, can include the following *issues*:
    - *Issue* (need): Put gas in the car.
    - *Issue* (wish): Call my daughter to say hello while driving.
    - *Issue* (obligation): Pay the phone bill.

- 🕐 The thinking center can process *issues* in all the segments of time:
    - 🕐 Present (interest): Listening to a song on the radio.
    - 🕐 Present (complaint): Having a toothache.
    - 🕐 Present (desire): Tasting the ice cream.
    - 🕐 Near future (obligation): Pay the phone bill.
    - 🕐 Near future (need): Go to lunch.
    - 🕐 Future (goal): Have a house or write a book.
    - 🕐 Near past (interest): Remember last Christmas dinner
    - 🕐 Immediate past (interest): Remember the phone number they just gave me.

    © Eduardo Padilla-Diaz - All Rights Reserved. epadilladiaz@gmail.com

Channels

① *"Stream useful information from sensory centers to thinkers and motor centers."*

16.1 Introduction

- ◆ Reflections on the *Channels*:
- ①◆ All sensory centers extract useful information from their sensory fields and pour it into different informatic streams called *Channels*.
- ①◆ The sensory centers stream useful information into the following *Channels*: *Reality*, *Thinker*, *Attention*, *Scene*, and *Motor*.

  1. ①◆ The *Reality Channel* streams all the information received by the sensors of a sensory center into a mental reality projector, allowing us to feel it continuously with excellent resolution.

  2. ①◆ The *Thinker Channel* streams to a *thinker* thinking messages produced by the object of interest under the focus, allowing us to understand and comprehend them.

     - ①◆ When an object of interest receives the focus from conscience, it becomes observed and entangled with thinking, awareness, and conscience. The focusing enables the emission of thinking messages through the *thinker* channel. The Entanglement chapter describes this mechanism in detail.

3. 🔵🔶 The *Attention Channel* is a generator of 'calls for attention.' When a sensory center detects a useful attention call from one of the objects of interest in its sensory field, it generates an attention call streamed through this *Channel*. The Thinking-2 chapter describes in detail how this *Channel* operates.

4. 🔵🔶 The *Scene Channel* maintains the list of objects of interest that have manifested in the sensory field, allowing us to locate them quickly and efficiently.

5. 🔵🔶 The *Motor Channel* streams commands to a motor center to automatically maintain the object of interest under focus with little or no intervention of thinking and consciousness.

   - 🔵🔶 The *Motor Channel* of a sensory center forms a *closed-loop controller* with a motor center to automatically maintain an object of interest under focus.
     - [closed-loop controller] wikipedia

© Eduardo Padilla-Diaz - All Rights Reserved. epadilladiaz@gmail.com

Entanglement

"How the object of interest under focus entangles with thinking and consciousness."

17.1 Introduction

- Reflections on *Entanglement* (part 1):
- Here, we'll explore *Entanglement* under the context of consciousness. Other *Entanglement* contexts are out of the scope of this exploration:
  1. The vocabularies are learning machines that help sensory centers extract objects of interest and their attributes from the sensory fields. (Book1.Vocabularies)
  2. The vocabularies are modal, so the Visual Center only has visual vocabularies; the Acoustic Center only has acoustic vocabularies; and the same thing happens with the other sensory centers. (Book1.Vocabularies)
  3. Remember that the innate visual vocabulary contains prototypes that allow us to recognize, in general, people, animals, plants, and things within the visual fields. (Book1.Vocabularies)
  4. Remember that the acquired visual vocabulary contains prototypes that allow us to recognize specifically our father, dog, tree, and house within the visual fields. (Book1.Vocabularies)
- When an object of interest receives the focus from the conscience, it becomes observed and entangled with the Thinking Center and conscience.

- ○ ◈ The focusing enables the emission of thinking messages from the sensory center to the Thinking Center. (Book2.Channels)

- ○ ◈ Enabling the emission of thinking messages bridges the information path between the entangled object and conscience. This path, or chain conformed by the entangled object, energy, detector, sensing, thinking, awareness, and conscience, is called *The Von Newman Chain*.
    - [John Von Neumann] wikipedia

- ◈ When an object of interest becomes entangled, the emission of its state is immediately enabled. The changes in its state are immediately reported to the Thinking Center for processing.

- ◈ When the conscience becomes aware of the object's state changes, it disentangles from the object and may focus on another object waiting for attention.

- ◈ In conclusion, the *Entanglement* of an object of interest under focus leads to the awareness of its state by the conscience.

© Eduardo Padilla-Diaz - All Rights Reserved. epadilladiaz@gmail.com

Quantumness

"The fabric of quantum thinking physics: a paradigm for modeling consciousness."

18.1 Introduction

- Reflections on *Quantumness* (Part 1):
- Here, we introduce a fundamental aspect of thinking and consciousness: They process discrete amounts of useful information in discrete amounts of time.
- We will not delve into quantum concepts at the subatomic level.
- We'll delve into quantum concepts, from sensations to awareness, under the realm of our mind's linguistic framework, where thinking and consciousness take place. I call this context *Quantum Thinking Physics*.
- Let's consider some fundamental concepts in the context of *Quantum Thinking Physics*:

1. *Quantumness*
    - *Quantumness* is a *space-time paradigm* the mind uses to efficiently handle the discrete information processing needed by consciousness' coherent comprehension of all thoughts and awareness superposition of all feelings.
        - [space-time] [paradigm] wikipedia
    - *Quantumness* is crucial for a series of hypotheses postulated in this work that help understand and model thinking, awareness, and consciousness.

2. *Quantum*
   - 🕐⊕ In abstract terms, a quantum is the smallest unit of something useful.

3. *Quantum Informatics*
   - 🕐⊕ Quantum informatics is the scientific study of processing discrete amounts of useful information in discrete amounts of time.

4. *Quantum Change of State*
   - 🕐⊕ A quantum change of state is an expression that embodies the changes in the properties of the entangled object of interest.
   - 🕐⊕ Remember that when an object of interest becomes entangled, the emission of its state changes is immediately enabled. The changes in its state are immediately reported to the thinking center for processing.

5. *Quantum Entanglement*
   - 🕐⊕ Quantum entanglement is a phenomenon that occurs after the object of interest under focus becomes entangled with conscience. This phenomenon lasts the time a *thinker* spends executing a finite-state machine step.
     - 🕐⊕ *Thinkers* are finite-state machines that process information in finite steps. They dismiss from conscience after each step and regain it under certain conditions to process the next step.
       - [finite state machines] wikipedia

6. *Slit*

    - ◐ A slit is a discrete window of conscious time where body matter is exposed to external energy.

7. *Quantum Present*

    - ◐ Quantum present is the amount of conscious time the *thinker* with focus spends processing a *finite-step*.

    - ◐ Quantum present is the amount of conscious time a *thinker* spends processing a *finite-step* of a quantum thought.

8. *Quantum Excitation*

    - ◐ A Quantum Excitation is the smallest linguistic unit that has meaning.

    - ◐ Quantum excitations are of different sizes.

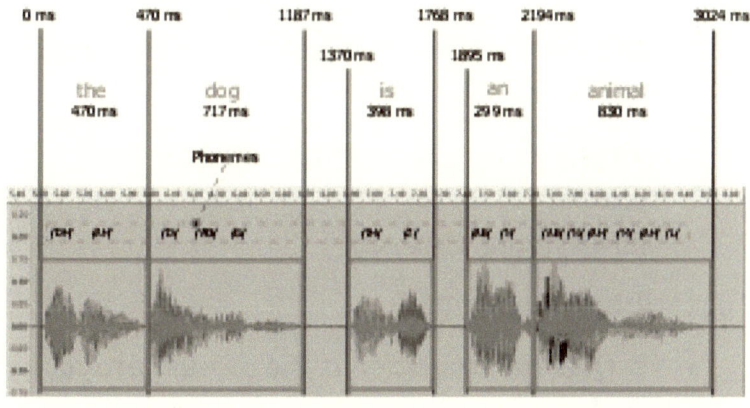

Fig. 18.1

- ◐ The figure graphically shows the sparsed acoustic excitations generated by an acoustic message composed of 5 words: <u>the</u> (470ms), <u>dog</u> (717ms), <u>is</u> (398ms), <u>an</u> (299ms), <u>animal</u> (830ms).

9. *Quantum Thought*

   - Quantum thought is a linguistic sentence that conforms with the mind's grammar; it is semantically understandable and comprehensible by the *thinker* processing it.
   - We will limit our exploration to sentences produced using visual and acoustic modalities. We will discuss sentences produced using other modalities later.
   - Visual and acoustic sentences can be *Declarative*, *Interrogative*, and *Imperative*:
     - A *Declarative* sentence (an affirmation) places a *thinker* in a learning state. For example, "The dog is an animal."
     - An *Interrogative* sentence (a question) places a *thinker* in a questioning state. For example, "Is the dog an animal?"
     - An *Imperative* sentence (a command, obedience) places a *thinker* in a command state — for example, "Silence!"

10. *Quantum Cognition*

    - Quantum cognition is the process of acquiring knowledge through quantum thoughts.
      - [Quantum Cognition] wikipedia

11. *Quantum Thinking*

    - Quantum thinking is an information fragmentation paradigm required to handle the concurrent processing of multiple *thinkers* efficiently. It is needed to fulfill consciousness' desire to process all outstanding issues in the thinking pool quickly and independently, according to their priority

- Quantum thinking is a paradigm that controls the ordered processing of multiple thoughts to guarantee isolation of conscious thought comprehension and coherence with their corresponding feelings awareness.

© Eduardo Padilla-Diaz - All Rights Reserved. epadilladiaz@gmail.com

Eduardo Padilla-Diaz

VC-2

◉ *"Visual Center 2: Automatic detection of object attributes, reflexes handling, and more."*

19.1 Introduction

- ◉ Reflections on the *Visual Center* (part 2 of 3):
- ◉ *Visual Center* (VC) is a sensory center of the body and mind, where everything related to visual informatics is exclusively processed. The *Visual Center* is explored in 3 parts: VC-1, VC-2, VC-P:
  - ◉ VC-1 explores the *Visual Center* at the photon and sensory levels. The first book of this series, The Rings of the Mind, explores it.
  - ◉ VC-2 explores the *Visual Center* at the operational level. The second book of this series, The Width of the Present, explores it.
  - ◉ VC-P explores the *Visual Center* at a functional level. This center gradually develops within the thinking models in each book.
- ◉ Here, we explore VC-2: the *Visual Center* at the operational level.
- ◉ Although the *Visual Center* operates autonomously mainly to give us the sensation of seeing, it also performs other equally important functions, like extracting useful information from the visual sensory signals. This useful information is poured into various informatic channels, which, under the control of the

attention, the conscience, and the Thinking Center, provide us with the understanding and comprehension of the visual experience so that we can reason to determine what to do in the immediate future.

- ❶❂ From an operational point of view, the *Visual Center* is composed mainly of the *eyes*, the *Visual Analysis Center*, the *Ocular Reflexes*, the *Ocular Motor Center*, the *Visual Vocabularies*, the objects of interest, the *Visual Attention Center*, and the *Informatic Channels*.

---

1. ❶❂ The *eyes* are two sensory organs that capture the visual information of the present from external space and transform it into a stream of visual reality, which is processed internally by the *Visual Analysis Center*.
    - [human eye] wikimedia

    - ❶❂ The figure on the right shows the internal diagram of a human eye. The eye captures light from external space. The cornea-1 and the crystalline lens-2 refract light on the retina-3, 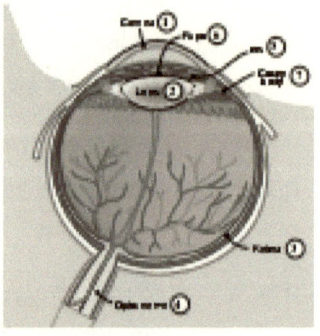 transducing the image into a parallel stream of electrical pulses, which are transported by the optic nerve-4 to the *Visual Analysis Center* for processing.
        - [eyes] [visual system] wikipedia

    - ❶❂ The *eyes* are positioned to form a stereographic capture unit. The information that each eye captures is simultaneously displaced by the distance that separates them, allowing the *Visual Center* to recreate the sensation of three dimensions (3D) obtained by combining the signals from both *eyes*.

- *The 1967 Nobel Prize in Physiology or Medicine was awarded jointly to Ragnar Granit, Haldan Keffer Hartline, and George Wald "for their discoveries concerning the primary physiological and chemical visual processes of the eye."*
  - [1967 Nobel Prize in Physiology-Medicine] nobelprize.org

2. The *Visual Analysis Center* receives the stream of visual reality from the *eyes*. It pours it unchanged into the projector of visual reality. As this information is received, it is processed automatically to obtain the following:

   a. The detection of objects of interest in the visual field based on matches with *Visual Vocabularies*.

   b. The detection of changes in the movement and direction of the objects of interest in the visual field is based on analyzing the integration of binocular visual information. David H. Hubel and Torsten N. Wiesel won the Nobel Prize in 1981 for this discovery.

   - *The 1981 Nobel Prize in Physiology or Medicine was divided, half awarded to Roger W. Sperry "for his discoveries on the functional specialization of the cerebral hemispheres," and the other half jointly to David H. Hubel and Torsten N. Wiesel "for their discoveries related to information processing in the visual system."*
     - [1981 Nobel Prize in Physiology/Medicine] nobelprize.org

   c. The detection of changes in color and shape of the objects of interest in the visual field.

   d. The detection of possible collision with objects of interest approaching us. If this is detected, an ocular reflex is executed so the body unconsciously protects itself from the potential impact.

e. ⓘ◉ The execution of closed-loop controllers used in *Ocular Reflexes*.

3. ⓘ◉ *Ocular Reflexes*: They are movements of automatic execution (unconscious), under the control of the *Visual Analysis Center* in concert with the moto-ocular system, in response to stimuli that do not receive or need help from conscious thought; and they are the photo-pupillary reflex, the optical convergence reflex, the corneal reflex, and the vestibular-ocular reflex.
    - [reflex] wikipedia

    a. ⓘ◉ The photo-pupillary reflex controls the pupil's diameter, helping the vision adaptation to various levels of light-darkness.
        - [papillary photo reflex] wikipedia

    b. ⓘ◉ The optical convergence reflex controls the pupil's constriction, the lens's accommodation, and convergence to achieve the best sharpness of the object we focus on.
        - [optical convergence reflex] wikipedia

    c. ⓘ◉ The corneal reflex is an involuntary blinking that protects the *eyes* when a foreign body stimulates the cornea. It provokes a response in both *eyes*. Flickering also occurs when sounds greater than 40 to 60 dB are produced.
        - [corneal reflex] wikipedia

    d. ⓘ◉ The vestibule-ocular reflex (VOR) stabilizes images on the retina during head movement, maintaining the gaze held in one place and producing eye movements opposite to the head movement.
        - [vestibulo-ocular reflex] wikipedia [vestibulo-ocular reflex] lumenlearning.com

4. ◉ The *Ocular Motor Center* is responsible for the voluntary and involuntary movement of the *eyes*, which helps to acquire, fixate, and track objects of interest.

   - ◉ The figure on the right shows the diagram of the *Ocular Motor Center*:
     - [human eye] wikimedia

   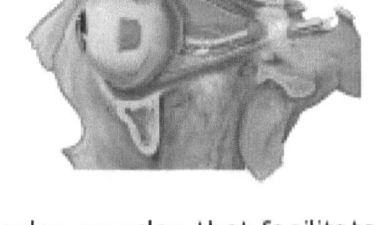

   - ◉ The *Ocular Motor Center* comprises six ocular muscles that facilitate eye movement and three cranial nerves that carry signals from the brain to control them.
     - [eye movement] wikimedia
   - ◉ The following figure shows all the eye movements that are achieved with the eye muscles:

   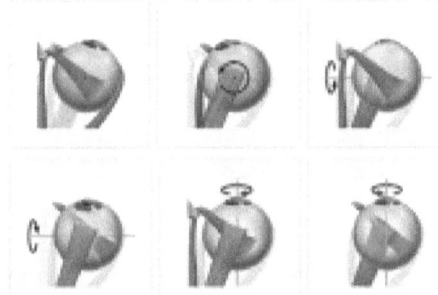

   Fig. 19.1

5. ◉ *Visual Vocabularies*: allow the *Visual Center* to detect and classify objects of interest in the visual fields. They are composed of innate and acquired *Visual Vocabularies*.

   a. ◉ The innate visual vocabulary allows the *Visual Center* to detect and classify, in a generic way, the four classes of objects of interest that can exist:

people, animals, plants, and things. It is genetic and necessary; without it, we could not operate or function as soon as we are born.

- Furthermore, for almost all species with *Visual Centers*, the innate visual vocabulary contains image prototypes, which allow each species to recognize its species within its visual field.

b. The acquired visual vocabulary allows the *Visual Center* to detect and classify objects of interest present in the visual fields in a specific way, for example:

- After birth, the acquired visual vocabulary allows us to recognize our mother, father, and other family members specifically.
- Sometime after birth, the acquired visual vocabulary allows us to recognize our habitat, other people, animals, plants, and other things we have seen specifically.

6. The objects of interest are those present in the visual field at a given moment that the *Visual Center* can recognize as part of its vocabularies.

- The objects of interest facilitate extracting the useful information they generate, manifested through space, time, and form attributes.
    - Space attributes, such as position and orientation.
    - Time attributes include their movement, direction in space, and whether they are moving away from or approaching us.

- Form attributes, such as size, color, and class. The class determines whether the oject is a person, animal, plant, or thing.

7. The *Visual Attention Center* works in concert with the conscience to determine which object of interest should have the focus. This center has two inputs and one output, which are connected to the *Visual Center*, and they are *Attention*, *Scene*, and *FocusReq*:

    - *Attention* is an input that determines the object of interest that should have focus. Still, the conscience makes the final decision based on a) the priority given to the attributes of the object of interest that calls for focus and b) the state of issues being processed by the *thinkers*.

    - *Scene* is an input that receives the list of objects of interest, which the *Visual Center* has detected in the visual field. It allows the conscience to locate them quickly and efficiently and determine which objects to pay attention to.

    - *FocusReq* is an output controlled by the conscience that tells the *Visual Center* what object of interest should be in focus based on what is reported by the *Visual Attention Center* and the interests of the conscience.

8. The *Informatic Channels*: All the useful information extracted in the *Visual Analysis Center* is poured into the following channels: Reality, Thinker, Attention, Scene, Motor, and *Ocular Reflexes*.

© Eduardo Padilla-Diaz - All Rights Reserved. epadilladiaz@gmail.com

Eduardo Padilla-Diaz

MC-2

⊕ *"Motor Center 2: Locomotion, balance, movement fuidity, precision, reflexes, command interface."*

20.1 Introduction

- ⊕ Reflections on the *motor centers* (part 2 of 2):
- ⊕ In part 1, we explored the *motor center* at the molecular, organelle, cellular, and muscular levels.
- ⊕ Here, we will explore the *motor center* at an operational level and a functional level:

20.2 Motor Center - Operational Level

- ⊕ The *motor center*, at an operational level, is composed of the following systems: *Locomotor, Cerebellum, Balance, Movement Fluidity, Precision, Reflexes,* and *Instincts*:

1. ⊕ The *Locomotor* system comprises the skeleton, skeletal muscles, cartilage, tendons, ligaments, joints, and other connective tissues. The skeleton 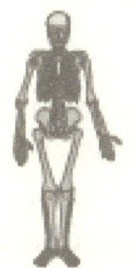  provides rigidity and shape to the body. The skeletal muscles hold bones in place and are responsible for

their movement; joints connect different bones. The cartilage prevents the ends of the bone from rubbing directly against each other. Muscles contract to move the bone attached to the joint.
- [locomotor system] wikipedia

2. ◈ The *Cerebellum*: At the operational level, it is one of the most essential signal-processing systems of the mind. It contains more than 50% of the total number of neurons in the brain; however, it represents approximately only 10% of its volume. It plays a vital role in motor control, in addition to being involved in cognitive functions such as attention and language, as well as emotional control; it does not initiate movement but contributes to coordination, precision, and synchronization. It receives information from sensory systems through thinking and other brain parts. It integrates these inputs to fine-tune motor activity.
    - [cerebellum] wikipedia

    - ◈ The figure on the right shows in red where the *Cerebellum* is located.

    - ◈ The following systems are part of the *Cerebellum*: *Balance*, *Movement Fluidity*, and *Precision*:

        - ◈ The *Balance* system maintains balance by adjusting body posture. It modulates motor commands to compensate for changes in body position or load on muscles. This system is entirely automatic and requires very little or no thinking assistance. It is located in the *Cerebellum*.
            - [cerebellum] neuroscience online

- ⊕ The *Movement Fluidity* system coordinates voluntary movements since most involve several muscle groups. Coordinate the timing and strength of these different muscle groups to produce fluidity in movements. This system is entirely automatic and requires very little or no thinking assistance. It is located in the *Cerebellum*.
    - [cerebellum] neuroscience online
- ⊕ The *Precision* system adapts motor learning to achieve precise movements. This semi-automatic system requires very little thinking assistance.
    - [cerebellum] neuroscience online

3. ⊕ *Reflexes* are automatic and almost instantaneous movements in response to a stimulus, which do not receive or need help from thinking. *Reflexes* are possible thanks to neural pathways called reflex arcs, which can act on an excitation before it reaches the brain. In humans, there are more than 50 types of *Reflexes* (see list of *Reflexes*), which can be generated in different parts of the body, for example:
    - [reflex] [list of reflexes] [reflex arc] wikipedia

    - ⊕ The pharyngeal reflex prevents objects in the oral cavity from entering the throat, except during normal ingestion. It also helps avoid choking and is a form of coughing.
        - [pharyngeal reflex] wikipedia
    - ⊕ The grasping reflex occurs in babies and causes their fingers to close when an object is placed in the palm of their hands.
        - [grasp reflex] wikipedia
    - ⊕ The sucking reflex occurs in babies. It is activated when the nipple is placed between their lips and

touches their palate —instinctively pressing it between the tongue and the palate to extract milk.
  - [sucking reflex] wikipedia

4. ⊕ *Instincts* are an inherent inclination of a living organism toward particular complex behaviors, containing both elements of our innate Knowledge and our acquired Knowledge. We will not consider *Instincts* in this adventure, but an example was given in the Knowledge chapter.
   - [instinct] wikipedia

20.3 Motor Center - Functional Level

- ⊕ Although the *motor center*, in general, is functionally composed of the following *motor centers* : the ocular, the vocal, the facial, the neck, the trunk, the upper extremities, the lower extremities, and others located within them. Here, we will treat them in a unified way, as the *motor center*, since they share much of what we will explain below.
- ⊕ At a functional level, each *motor center* comprises a *Command Interface*, *Cerebellum*, and *Muscle Groups*.
  1. ⊕ The *Command Interface* is the component of a *motor center* that receives commands from the Thinking Center. All voluntary movement in the body is controlled primarily by the Thinking Center. The Thinking Center sends high-level motor commands to the command interface, which are interpreted and converted into thousands of low-level motor commands

(microinstructions), with the help of the *Cerebellum*, which is sent to the different muscle groups so that they are executed efficiently and gracefully.
  - [mensajes pensantes] [pensamiento]

2. ⊕ The *Cerebellum*, at a motor functional level, receives motor commands and is involved in the following functions:
   - [Cerebellum] neuroscience online [Cerebellum] thebrain.mcgill.ca

   - ⊕ Modifies motor commands so that movements are more adaptive and precise.
   - ⊕ Modulates motor commands to compensate for changes in body position or changes in load on muscles.
   - ⊕ Coordinates voluntary movements since most movements involve several muscle groups. The *Cerebellum* coordinates the timing and strength of these different muscle groups to produce fluidity in movements.
   - ⊕ Adapt and adjust motor learning to achieve more precise movements.
   - ⊕ Routes motor commands to the *motor centers* that are needed.

3. ⊕ *Muscle Groups*: Most muscles are part of groups that work together to fulfill a specific body function; for example, there are groups in the hand, foot, tongue, and the extraocular muscles of the eye. These groups receive operative commands from the *Cerebellum* to satisfy the Thinking Center's functional commands.
   - [Skeletal muscle : groups] wikipedia

© Eduardo Padilla-Diaz - All Rights Reserved. epadilladiaz@gmail.com

Eduardo Padilla-Diaz

Thinking Messages

◐ *"Sensory informatics paradigm bridges changes and states to the thinking system."*

21.1 Introduction

- ◆ Reflections on *Thinking Messages* (part 2):
- ◐◆ *Thinking Messages* were introduced in the first book, The Rings of the Mind, in the following chapters: Objects of Interest, Fields, and Thinking-1. Here, we propose a more advanced version.
- ◐◆ *Thinking Messages* are a functional-level paradigm that meets the necessary informatics requirements so that the sensory centers can efficiently communicate to the brain's thinking system the state and changes of the objects of interest captured by the body's sensory centers.
- ◐◆ *Thinking Messages* are made of *Infogram* objects that facilitate the transport and processing of informatic expressions between sensory, thinking, and knowledge centers.
- ◐◆ An *infogram* is functionally similar to a mail package used to exchange and transfer informatic expressions.
    1. ◐◆ An *infogram* contains three components: *Label*, *Content*, and possibly a *State*:
        - ◐◆ *Label*: indicates the destination where the package is sent, and the recipient is added to the package at the time of shipping.

- *Content*: represents an expression that includes changes or new attributes about the object of interest under focus.
  - For example, in the visual context, the *Content* expression could contain information about the position, distance, orientation, size, and posture of the object of interest under focus.
  - Another example is in the auditory context, where the *Content* expression could contain information about the position, distance, orientation, and intensity of the object of interest under focus.
- The *State* is an expression that provides access to the sensory, semantic, and scenic information we currently have about the object of interest under focus so that the Thinking Center can make a better decision about the current changes in the object attributes.
  a. Let's consider that transporting a sensory expression through the body's networks would be impossible since it would entail excessive consumption of the networks' capacity and energy and considerably degrade the thinking system response time. That is why sensory expressions are stored in the memory of the sensory center where they were produced. Instead, a sensory token is used to transport it efficiently across networks, and at the receiving end, it is used to access the sensory expression remotely.
  b. The sensory token represents a link to the sensory expression, which stores the sensation's

capture in sensory code so that it can be reproduced virtually in the future with the help of imagination.

   c. 🔹 Transporting the semantic and scenic expressions would entail even worse degradation of the network capacity and its response time. That is why referential tokens are used to transport efficiently semantic and scenic representations and to provide efficient remote access to their information.

   d. 🔹 The semantic token represents a link to our semantic knowledge of the object of interest under focus.

   e. 🔹 The scene token represents a link to the scene expression, which stores the scene's capture in sensory code so that it can be reproduced virtually in the future with the help of imagination.

2. 🔹 Note: At the operational level, the way *Infograms* are assembled in sensory centers is outside the scope of this exploration, but it has to be a very fast process (on the order of 10 milliseconds) since all its components are references to already pre-computed data.

3. 🔹 *Infograms* at the sensory level are modal. There is a type of *Infogram* for every modal context. Here, we will limit the discussion to the following contexts: Visual, Acoustic, and Motor.

   - 🔹 Visual: It uses exclusively visual *Infograms* called photograms generated by the visual center.

- Acoustic: It uses exclusively acoustic *Infograms* called audiograms generated by the acoustic center.
- Motor: It uses motor *Infograms* exclusively, called motograms, generated by the motor center and the thinking system.

4. All types of *Infograms* have the following characteristics:

- All *Infograms* are valid at a thinking level, meaning the thinking system can process any *Infogram*. However, the thinking system only processes content and semantic expressions, which are strictly linguistic. From sequences of *Thinking Messages*, linguistic sentences are extracted, and the thinking system transforms them so that the consciousness understands and comprehends them semantically.
- All *Infograms* are valid at a cognitive level in the knowledge context.
- *Infograms* at the sensory level are only valid at the sensory center that produces or reproduces them.

© Eduardo Padilla-Diaz - All Rights Reserved. epadilladiaz@gmail.com

Thinking-2

① *"Quantum thinking in all thinkers, attention arbitration, anticipation, and more."*

22.1 Introduction

- ◉ Reflections on thinking (part 2):
- ①◉ At this stop, a new P-2 model presents how thinking would operate from an animal perspective.
  - ①◉ We will cover a specific case of a wild animal species in which a lion cub follows his mother along a path to a water source, where they will quench their thirst.
  - ①◉ This case demonstrates how animals have the physical, virtual, motor, emotional, and affective capacity to operate successfully in a conducive environment without requiring the ability to communicate with words.
- ①◉ But first, we are going to consider certain concepts, which are fundamental in the functionality of us as animals, and which are *Prediction*, *Search*, and *Affection*:

  1. ①◉ *Prediction*: allows us to anticipate the immediate future with great certainty and compensate for the delay caused by the *width of the present*, which, as we saw previously, is the time interval that exists between the exposure to reality at a sensory center and the delay in consciously perceiving it by its respective

center. *Prediction* makes us feel reality without the delay caused by the *width of the present*. To illustrate *Prediction*, we are going to describe how we can walk in a semi-automatic way, almost without realizing it:

a. 🔅 The Visual Center is capable of automatically detecting changes in the attributes of an object of interest, such as changes in its position.

b. 🔅 The visual imagination comes into play, allowing the Visual Center to virtually see the position of a moving object of interest in the immediate future, enabling us to anticipate its position in the virtual future.

- 🔅 "Consider catching a ball. It takes several tens of milliseconds for information from the eye to reach the brain and about 120 milliseconds before we can take action based on that information. During this time, the ball continues to move, so the brain's information about where the ball is will always lag where the ball actually is. In sports such as tennis, cricket, and baseball, balls travel at speeds well more than 100 km per hour, which means the ball can move more than 3 meters during this time delay. If we perceived the ball's position based on the most recent information available to the brain, we would never be able to catch or hit it with any accuracy."
  - [predicting-the-present] [unconscious-visual-biases] theconversation.com [predictions] pnas.org
- 🔅 The photo (Shutterstock/Herbert Kratky) shows how a soccer goalkeeper can catch a ball that changes position up to 3 meters within his thinking

cycle. Thanks to his visual imagination, he can anticipate its position in the immediate future.

c. 🌓 This ability to anticipate, added to the ability to learn and use semi-automatic motor procedures, allows an animal to walk on a path semi-automatically, under the control of the subconscious and with very little conscious supervision.

d. 🌓 Reflexes come to the rescue to overcome possible unexpected changes in what is anticipated. For example, suppose another animal unexpectedly crosses an animal's path. In that case, the Visual Center automatically generates a reflex that causes the animal to avoid a collision. Through a call from the Visual Center to the center of attention, they make the conscience finally resolve the unexpected.
   - [reflex] wikipedia

e. 🌓 Thus, the automatic detection of position changes, visual imagination, automatic prediction (anticipation) of the position of an object of interest in the immediate future, the subconscious, and reflexes make it possible for us to walk without realizing it; that is, semi-automatically with the help of the subconscience, and only in unexpected cases, with the help of the conscience.

f. 🌓 In this exploration, the term 'walk' refers generically to the ability of a body-mind to move over terrain. Mechanically, human steps differ from those of a quadruped animal, such as a cub. In both cases, this capacity is assumed to require minimal conscious intervention.

2. ①◉ *Search* refers to finding the meaning of the sensory excitations in the innate and acquired semantic knowledge. Based on what has been described in Book1 about knowledge, vocabularies, objects of interest, and prototypes, we can state the following:

   a. ①◉ The cub's Visual Center has the prototype of its species in its innate visual vocabulary, allowing it to find any animal of its species among the objects of interest in its visual field.

   b. ①◉ In its acquired visual vocabulary, the cub's Visual Center has the prototype of the mother, allowing it to find her specifically among the objects of interest in its visual field.

   c. ①◉ The cub's Smelling Center has the mother's prototype in its prenatally acquired smelling vocabulary, allowing it to find her specifically among the objects of interest in its smelling field; this is why animals can recognize their mothers immediately after birth through smell.

      ▪ ◉ Note: The Smelling Center is not included in the P-2 model.

   d. ①◉ The cub's Acoustic Center has the mother's prototype in its prenatally acquired acoustic vocabulary, allowing it to find her specifically among the objects of interest in its acoustic field; this is why animals can acoustically recognize their mother immediately after birth.

- - ⊕ Note: The Acoustic Center will not be included in the P-2 model.

3. ⊕ *Affection*, in this context, refers to the emotional bond that binds animals of the same family.

    a. ⊕ The cub's and mother's emotional centers create a powerful emotional bond that manifests itself with certain biochemical rewards that affect mutual behavior.

        - ⊕ Note: The Emotional Center is not covered in this adventure, but I mention it to introduce its importance in mutual behavior between animals and humans.

    b. ⊕ The sum of the visual, smelling, acoustic, and emotional effects means that cubs can easily follow their mothers. This sum is called a Sigma and it is covered in the next book: Tokens and Words.

- ⊕ With these considerations, we can evolve the previous P-1 model presented in the first book of this series. We are going to add a Thinking Center, improve *thinkers*, improve the Motor Center, add a Visual Attention Center, improve the Visual Center at an operational level, and guarantee the cognitive coherence of the Visual Center:

1. ⊕ We will add a new Thinking Center that allows us to have several *thinkers* and process several thoughts simultaneously.

    a. ⊕ To illustrate the new model, we will add three *thinkers*: *Thinker* 1 will help the cub follow its mother; *Thinker* 2 will help the cub follow the path; and *Thinker* 3 will allow the cub to walk the path.

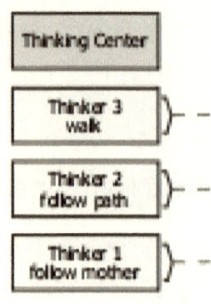

b. ①❖ This case covers three issues: follow the mother, follow the path, and walk.

- ①❖ Remember that an issue is a need, complaint, desire, interest, obligation, goal, or something being resolved, must be resolved, is being resolved, or has already been resolved.

- ①❖ Issues are critical because they are the essence of thoughts, and each needs a *thinker* to process them.

2. ①❖ We are going to improve the *thinkers* so that they are more effective in the time domain:

   a. ①❖ We will change the *thinkers* so that they do not spend time unnecessarily interrogating to see if they have received a new message ('polling'). Instead, they will remain suspended and automatically activated when a new message arrives (*event-driven*). These changes make thinking cycles more responsive and more effective.
      - [event-driven] wikipedia

   b. ①❖ The human visual system processes 10 to 12 images per second. Li, H., Wang, Y., & Wang, L. (2012). "A review of non-contact, low-cost physiological information measurement based on photoplethysmographic imaging."
      - [physiological information measurement] doi.org [visual system images per second] wikipedia

- 🔹 Based on this, we will assume that each thinking cycle lasts 80 to 100 milliseconds to effectively manage the images that the visual system generates and prevent useful information from being lost.

    c. 🔹 Note: A thinking cycle is the time a *thinker* spends processing a thinking message from a sensory center.

3. 🔹 Improvements in the Motor Center: It must be able to process several motor functions simultaneously and independently. We can move our heads and eyes and, in turn, our hands and fingers, and at the same time, our feet and so on. We can do all these movements in a coordinated, harmonious, and concurrent way. To begin to achieve this, we are going to divide the Motor Center into two large functional centers: the Head Motor Center and the Body Motor Center:

    a. 🔹 The Body Motor Center comprises several operational motor functions responsible for moving most body parts, such as the arms, hands, fingers, legs, feet, and others, semi-automatically with minimal conscious intervention. Those operational functions are combined to implement several issue-driven motor functions like walking, taking, and more.

        - 🔹 For now, we will only consider the walking motor function, which we will call the Walker.

    b. 🔹 The Head Motor Center comprises the visual, vocal, auditory, facial, and neck motor centers. All of them carry out their functions semi-automatically with minimal conscious intervention. Those operational functions are combined to implement several issue-driven motor functions, like observing a moving object.

- ⓘ⊕ For now, we will only consider combining the visual and neck motor functions to track a moving target. We will call this motor function the Tracker.

c. ⓘ⊕ The following figure shows a partial mental diagram of the new Motor Center and its association with the conscience, the Visual Center, and the Thinking Center.

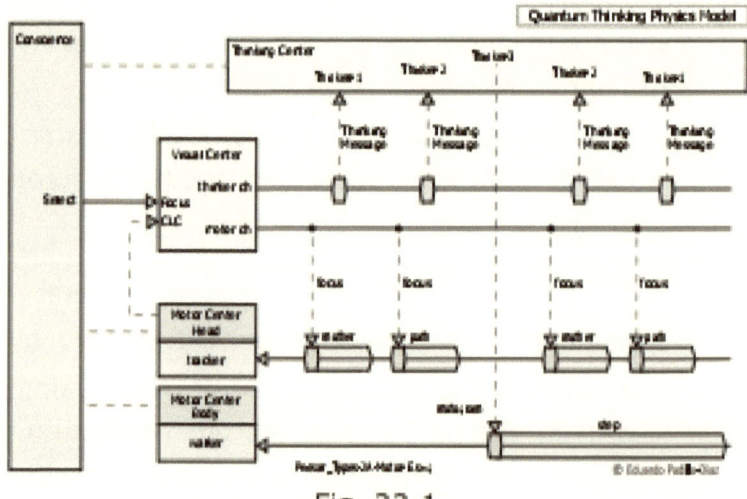

Fig. 22.1

d. ⓘ⊕ The Motor center forms a *closed-loop controller* with the Visual Center to automatically maintain the object of interest under focus with little or no intervention from thinking and consciousness.
- [closed-loop controller] wikipedia

e. ⓘ⊕ The Tracker is under the command of the Visual Center, which is under the control of the conscience, to move the head and eyes and thus be able to observe the mother and then observe the path.

f. 🔹 The Walker is under the command of *Thinker* 3; operationally, it is in charge of walking the path according to the results of the observations about the mother and the path. Functionally, it is responsible for walking the path without stumbling, according to the speed with which the mother walks, staying as close to her as possible.

g. 🔹 Observations about the mother: The conscience commands the Visual Center to focus on the mother; this causes *Thinker* 1 to begin receiving the thinking messages coming from the mother, and from them, he obtains: 1) the distance to the mother; 2) the relative speed with which the mother walks; 3) any communicative gesture that the mother makes to the cub.

h. 🔹 Observations on the path: Then the conscience commands the Visual Center to focus on the path; this causes *Thinker* 1 to begin receiving the thinking messages coming from the path, and from them, he obtains: 1) whether the direction needs to be adjusted: 2) if there is an obstacle, adjust the distance of the step.

4. 🔹 We will add the Visual Attention Center so that, under the control of the conscience, they can determine which object of interest should have the focus and which

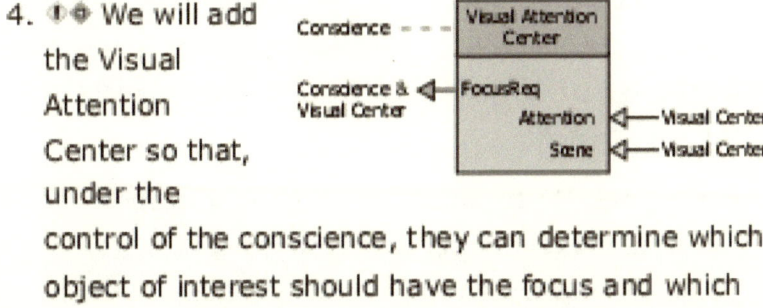

thinker should process it. This center has two inputs and one output, which are connected to the Visual Center and are called *Attention*, *Scene*, and *FocusReq*:

a. *Attention*: It is an input that determines the object of interest that should be in focus. Still, the conscience makes the final decision based on a) the priority given to the attributes of the object of interest that calls for focus and b) the state of issues being processed by the *thinkers*.

b. *Scene*: It is an input that receives the list of objects of interest, which the Visual Center has detected in the visual field. It allows the conscience to determine which objects it should pay attention to and on which object to place the focus.

- Each object of interest on the scene list is already classified by its type —human, animal, plant, thing— and by its position in space, which makes it easier for the Visual Center to find it quickly when the conscience asks it to focus on it.

- However, factors such as familiarity, attraction, dangerousness, etc., determine the degree of interest in an object.

c. *FocusReq* is an output controlled by the conscience that tells the *Visual Center* what object of interest should be in focus based on what is reported by the *Visual Attention Center* and the interests of the conscience.

- Note: When an object of interest is observed, it is because it is under focus, which causes its thinking messages to be enabled; these, in turn, go to the Thinking Center to awaken the *thinker* in charge (if he is not already in that state), so it can process them.

5. ◐◉ We are going to improve the Visual Center at the operational level to facilitate its interface with the conscience, the Visual Attention Center, and the Head Motor Center:

   a. ◐◉ The following figure shows a partial mental diagram of the new Visual Center, controlled by the Visual Attention Center:

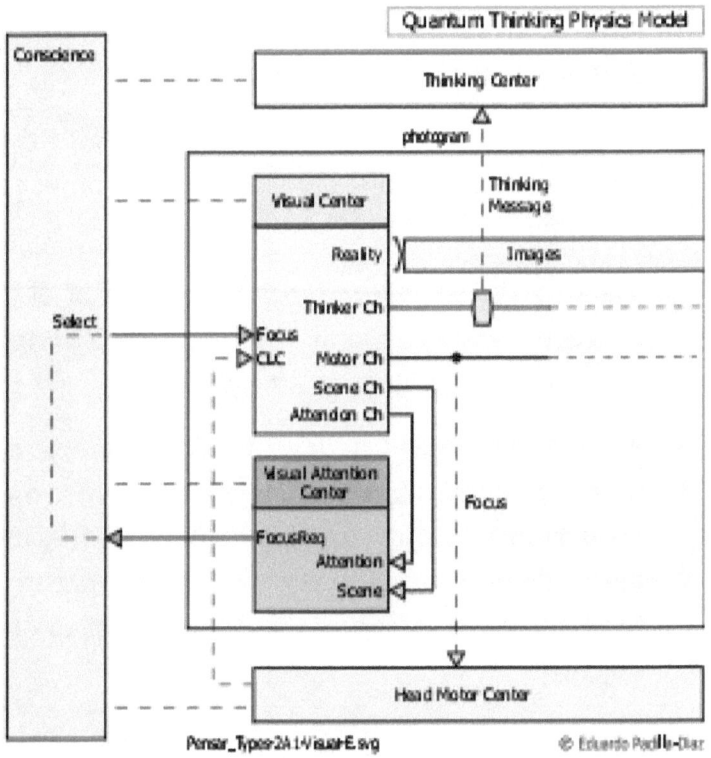

Fig. 22.2

   b. ◐◉ The *Thinker Channel* streams to a *thinker* thinking messages produced by the object of interest under the focus, allowing us to understand and comprehend them.

- 🛈◈ When an object of interest receives the focus from conscience, it becomes observed and entangled with thinking, awareness, and conscience. The focus enables the emission of thinking messages through the *Thinker Channel*. The Entanglement chapter describes this mechanism in detail.

c. 🛈◈ The *Motor Channel* streams commands to a motor center to automatically maintain the object of interest under focus with little or no intervention of thinking and consciousness.

- 🛈◈ It forms a *closed-loop controller* with the Visual Center to automatically maintain the object of interest under focus.

d. 🛈◈ The *Scene Channel* maintains a list of objects of interest that have manifested visually, allowing us to locate them quickly and efficiently.

e. 🛈◈ The *Attention Channel*: It is a generator of 'attention calls.' When an object of interest manifests visually in a useful way, this channel immediately informs the Visual Attention Center so that it, with the help of the conscience, decides what to do with the call.

f. 🛈◈ *Focus*: It is an input from the Visual Center, controlled directly by the conscience and indirectly by the *FocusReq* output of the Visual Attention Center, that can be modified or taken as such by the conscience.

- The scene's list of objects with their respective positions is automatically refreshed unconsciously by the Visual Center; thus, when the conscience instructs the Visual Center to focus on a specific object of interest, the Visual Center already knows its location and directly commands the Head Motor Center to position the head and eyes on it.

- Again, when an object of interest is observed, it is because it is under focus, and this causes its thinking messages to be enabled; these, in turn, go to the Thinking Center to awaken the *thinker* in charge (if he is not already in that state), so it can process them.

6. We will improve the Visual Center at a cognitive level, compensating for inevitable delays.

    - There is a delay between the moment information is extracted from the object of interest that has the focus and the moment in which decisions are made in the Thinking Center. We will assume that this delay is on the order of 100 milliseconds; this means that when decisions are made in the Thinking Center, 100 milliseconds have already passed. If the object of interest is moving, it will be somewhere else. Therefore, the *thinker* will be interacting in the immediate past, where the object is no longer, and not in the present of reality, where the object is.

    - Let's see how this delay is compensated:

        - Instead of using the position of the object of interest based on the immediate past—which is how we have been doing it until now — its position is anticipated in the immediate future, extrapolating the previous position with the extracted position and extending it 100 milliseconds into the future. That supposed new position is what is included in the thinking message sent to the Thinking Center.

- 🕐 The Visual Center can anticipate with certainty the movement of the object of interest under focus as long as its change in position is within the visual field.

7. 🕐 The figure shows the final mental diagram of the P-2 model, described in detail through this exploration.

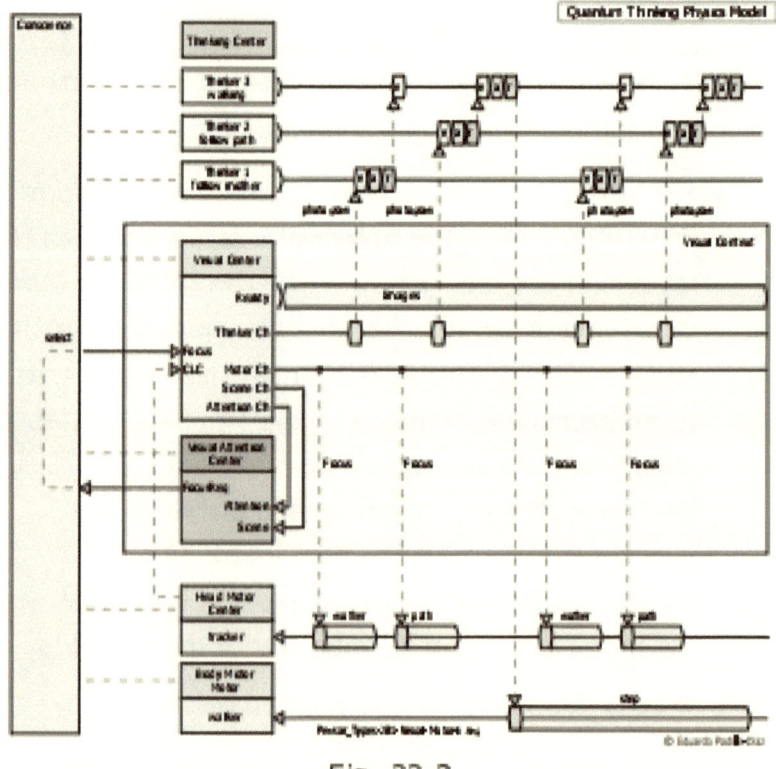

Fig. 22.3

© Eduardo Padilla-Diaz - All Rights Reserved. epadilladiaz@gmail.com

Awareness

⊕ *"Is the state of conscious comprehension of all thinking and feeling of all sensations."*

23.1 Introduction

- ⊕ Reflections on *Awareness* (part 1):
- ⊕ In philosophy and psychology, *Awareness* is a concept about knowing, perceiving, and being cognizant of events.
  - [awereness] wikipedia
- ⊕ Here, *Awareness* is also the sensing of our conscious comprehension of all thinking and the complementary conscious feeling of all sensations.
  - ⊕ The thinking encompasses the aggregate conscious comprehension of all *thinker* issues being resolved in the attention pools.
    - ⊕ The attention pools are a conglomerate of all the attention centers. The Visual Attention Center has already been covered. Still, we will explore the attention centers for the acoustics, smelling, and emotions later.
  - ⊕ The sensations encompass the aggregate conscious feeling of all sensations in the physical and virtual pools of reality, emotions, attractions, associated experiences, associated expectations, and motor activity.
- ⊕ Here, we'll add to the latest thinking model interaction with consciousness to give us a sense of

Eduardo Padilla-Diaz

Awareness in the following contexts: *Thinking*, *Visual*, and *Motor*:

1. 🕮 *Thinking* encompasses the following issue-driven *Awareness*: *Thinker*-1, *Thinker*-2, and *Thinker*-3.

    a. 🕮 *Thinker*-1 *mother awareness*: When the cub places the focus on the mother, her position is automatically tracked by the cub's *head motor center*, and that entangles the mother state with the visual sensory system; the thinking messages are enabled and sent to *Thinker*-1 for processing; the mother awareness is informed of the mother state, and from it, the cub obtains: 1) the distance to the mother; 2) the relative speed with which the mother walks; 3) any communicative gesture that the mother makes to the cub.

    b. 🕮 *Thinker*-2 *path awareness*: When the cub places the focus on the path, its position is automatically tracked by the cub's *head motor center*, and that entangles the path state with the visual sensory system; the thinking messages are enabled and sent to *Thinker*-2 for processing; the path awareness is informed of the path state, and from it, the cub obtains: 1) whether the direction needs to be adjusted: 2) if there is an obstacle, adjust the distance of the step.

    c. 🕮 *Thinker*-3 *walking awareness*: When the cub focuses on walking, it is handled mainly by the *body motor center*. This center is responsible for moving most body parts, such as the arms, hands, fingers, legs, feet, and others, semi-automatically with minimal conscious intervention. Those operational functions are combined to implement several issue-driven motor functions, such as walking.

- Walking is under the control of *Thinker* 3, operationally and functionally:
  - Operationally, it is in charge of walking the path.
    - According to the awareness of the mother issue, the cub knows 1) the distance to the mother and 2) the relative speed with which the mother walks.
    - According to awareness of the path issue, the cub knows 1) whether the direction needs to be adjusted and 2) if there is an obstacle, adjust the step's distance.
  - Functionally, it is responsible for walking the path without stumbling, according to the speed with which the mother walks, staying as close to her as possible.

2. *Visual Awareness* encompasses the following sensations: *Reality, Thinker, Scene, Head Motor,* and *Attention*:

    a. *Reality Awareness*: When the reality channel streams all the information received by the Visual Center's sensors into the mental reality projector, *Reality Awareness* is informed of the reality state, allowing the cub to see it with excellent resolution.

    b. *Thinker Awareness*: When a visual object of interest receives the focus from conscience, it becomes observed and entangled with thinking, awareness, and conscience. The focusing enables the emission of thinking messages through the visual *Thinker* Channel. The *Thinker's Awareness* informs the conscience of which *thinker* is handling the state of the visual object under focus.

c. ①� *Scene Awareness*: The Visual Center automatically refreshes the scene state unconsciously. *Scene Awareness* informs the conscience about the scene's list of visual objects of interest, its type: human, animal, plant, thing, and its position in space.

d. ①� *Head Motor Awareness*: When a visual object of interest receives the focus from conscience, it becomes observed and entangled with thinking, awareness, and conscience. *Head Motor Awareness* informs the conscience about how well the object is automatically tracked.

e. ①� *Attention Awareness*: Our conscience can rotate the focus among several issues it keeps on the *issue's attention list*. The priority of each issue in the list can vary from moment to moment and is given by the conscience according to the results of the last perception. *Attention Awareness* informs the conscience about which visual issue is being served.

3. ①� *Motor Awareness*: Applies to Head and Body Motor Centers and encompasses *Intention*, *Monitoring*, *Abilities*, and *Deficits*:

   a. ①� *Intention* informs the conscience of the desire to perform an action. For example, to execute a step.

   b. ①� *Monitoring* informs the conscience of the state of execution of the action and identifies motor errors.

   c. ①�*Abilities* inform the conscience of the capacity to recognize the general state of our motor abilities about the performance of execution of specific actions.

   d. ①�*Deficits* inform the conscience of the capacity to recognize the general state of the consequences of motor impairment.

- 🕐 The following figure shows the latest mental diagram of the P-2 model, described in detail through this exploration. Significant changes have been made: the addition of the thinking context container, visual context container, motor context container, and all the awareness connections of these contexts with conscience.

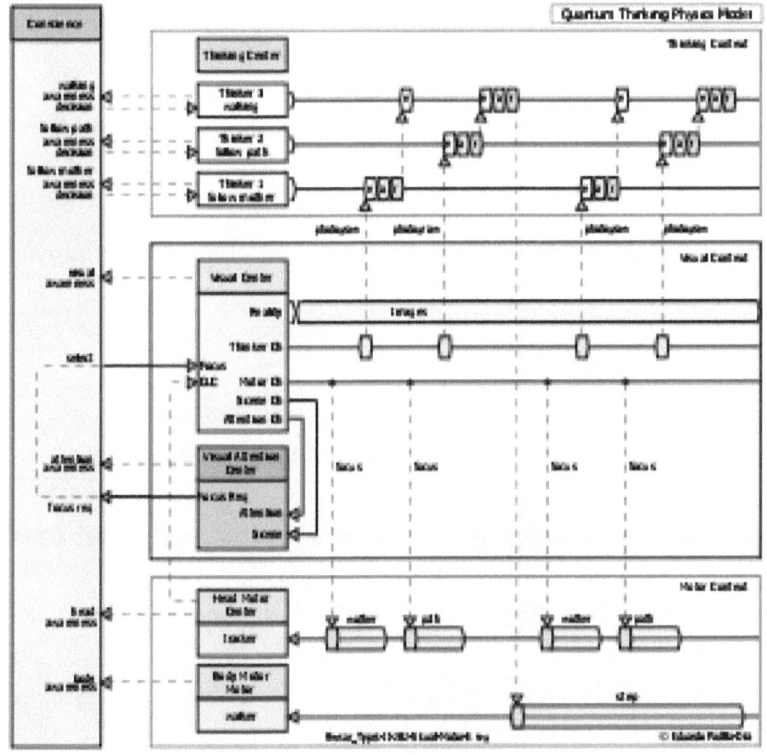

Fig. 23.1

© Eduardo Padilla-Diaz - All Rights Reserved. epadilladiaz@gmail.com

Epilog 2

- Thematics of the next book in the series: Tokens and Words.

0. [Cover-3] *Tokens and Words - Book 3 of the series: Modeling the Body-Mind.*
1. [TOC] *Thematics and Exploration Plan.*
2. [Conventions] *Considerations about revisions, references, TOCs, and Epilogs.*
3. [Abstract 3] *Neuroinformatics and linguistics are behind understanding and comprehension.*
4. [Tokens and Words] *The words we read are just tokens representing meaning.*
5. [Learning] *Introduction to our conscious ability to learn.*
6. [Syntax and Grammar] *Brief introduction to the rules that underlie human communication.*
7. [Events and Sigmas] *Store in knowledge states of quantum thinking that can be recalled.*
8. [Memories-3] *Semantic, episodic, procedural memories.*
9. [Habitat] *Only in these propitious places does the body-mind flourish and endure.*
10. [Stages and Scenes] *Spaces where quantum sensing takes place and quantum thinking sparks.*
11. [Automation] *Encompass the learning and automation of useful motor functions.*
12. [Mentalization] *Interface between sensory physiology and mental psychology.*
13. [Monitoring] *Guarantees the integrity and proper functioning of our body-mind.*
14. [Neuroinformatics] *How neural signal information is encoded and decoded.*
15. [Slits] *Discrete windows of conscious time where body matter is exposed to external energy.*
16. [Quantum Cognition] *Is the process of acquiring knowledge through quantum thoughts.*

17. [Acoustic Center] *Sensory center where acoustic information is acquired and processed.*
18. [Attention Pools] *Conscience assistants that help arbitrate all thinking issues priorities.*
19. [Finite State Machines] *Machines specialized in handling sparsed sequential information.*
20. [Thinking-3] *Added automatas, reading road signs, GPS listening, driving a car to work.*
21. [Awareness-2] *Includes acoustics awareness, attention pools, and more.*
22. [Imagination] *Virtuality transformer to cross the barriers of space-time, matter, gravity, and +.*
23. [Intelligence] *Inventions are created when solutions do not exist, and the need can not wait.*
24. [Epilog-3] *Thematics of the next book in the series: The Domains of the Conscience.*
25. [Copyrights] *List of copyright certificates that protect these works.*
26. [Dedication] *To my children: Mariana, Andres, and Camilo.*
27. [Back-cover-3] *Brief about the author Eduardo Padilla-Diaz and his works.*

© Eduardo Padilla-Diaz - All Rights Reserved. epadilladiaz@gmail.com

Eduardo Padilla-Diaz

Copyrights

ℹ️ List of copyright certificates that protect these works:

Entity	Title	Copyright Certificate	Date
Colombian Ministry of Interior	Ancho del Presente	10-1142-210	Jul 1, 2023
Colombian Ministry of Interior	Anillos de la Mente	10-1152-391	Aug 10, 2023
Colombian Ministry of Interior	Ancho del Presente	10-1170-247	Oct 24, 2023
Colombian Ministry of Interior	Rings of the Mind	10-1178-253	Nov 30, 2023
Colombian Ministry of Interior	Awareness	10-1204-369	Apr 10, 2024
Colombian Ministry of Interior	Quantumness	10-1204-416	Apr 10, 2024
Colombian Ministry of Interior	Entanglement	10-1205-1	Apr 10, 2024
Colombian Ministry of Interior	Time Segments	10-1205-2	Apr 10, 2024

Eduardo Padilla-Diaz

Dedication

To my children: Mariana, Andres, and Camilo.

Eduardo Padilla-Diaz

The Width of the Present

Eduardo Padilla-Diaz

The Width of the Present

www.ingramcontent.com/pod-product-compliance
Lightning Source LLC
Chambersburg PA
CBHW031920240526
45464CB00021B/615